Energy Science, Engineering and Technology

Energy Science, Engineering and Technology

The Future of Biodiesel
Michael F. Simpson (Editor)
2022 ISBN: 979-8-88697-166-8 (Softcover)
2022 ISBN: 979-8-88697-172-9 (eBook)

Nanotechnology Applications in Green Energy Systems
Rajan Kumar, PhD and Tangellapalli Srinivas, PhD (Editors)
2021 ISBN: 978-1-68507-451-7 (Hardcover)
2022 ISBN: 978-1-68507-479-1 (eBook)

Energy Conversion Systems: An Overview
Sanjeevikumar Padmanaban, PhD
and Saurabh Mani Tripathi, PhD (Editors)
2021 ISBN: 978-1-53619-131-8 (Hardcover)
2021 ISBN: 978-1-53619-200-1 (eBook)

Advanced Power Systems and Security: Computer Aided Design
Muna Hamid Fayyadh and Samir Ibrahim Abood
2020 ISBN: 978-1-53618-785-4 (Hardcover)
2020 ISBN: 978-1-53618-863-9 (eBook)

Energy Storage Systems: An Introduction
Dr. Satyender Singh (Editor)
2020 ISBN: 978-1-53618-873-8 (Hardcover)
2020 ISBN: 978-1-53618-910-0 (eBook)

More information about this series can be found at https://novapublishers.com/product-category/series/energy-science-engineering-and-technology/.

Eugene Chaikovskaya

Smart Grid Technologies in Electric Systems for Renewable Energy

DOI: https://doi.org/10.52305/KJMW9889

NOTICE TO THE READER

Additional color graphics may be available in the e-book version of this book.

Library of Congress Cataloging-in-Publication Data

Names: Chaikovskaya, Eugene
Title: Smart grid technologies in electric systems for renewable energy / Eugene Chaikovskaya.
Description: New York : Nova Science Publishers, [2022] | Series: Energy science, engineering and technology | Includes bibliographical references and index. |
Identifiers: LCCN 2022051920 (print) | LCCN 2022051921 (ebook) | ISBN 9798886973877 (paperback) | ISBN 9798886974324 (adobe pdf)
Subjects: LCSH: Smart power grids.
Classification: LCC TK3105 .S5328 2022 (print) | LCC TK3105 (ebook) | DDC 621.042--dc23/eng/20221223
LC record available at https://lccn.loc.gov/2022051920
LC ebook record available at https://lccn.loc.gov/2022051921

Published by Nova Science Publishers, Inc. † New York

Contents

Introduction

The book includes six chapters in which represented integrated Smart Grid systems for coordination of production and consumption of the electric power on the basis of forecasting of change of parameters of technological processes are developed. Advancing decisions to support the functioning of electrical systems is based on a mathematical description of the architecture of technological systems, methodology of mathematical description of the dynamics of energy systems, causal graph method, mathematical description Smart Grid of maintaining the functioning of electric systems.

Chapter 1 – The mathematical substantiation of maintenance of the operation of the technological systems Smart Grid based on the architecture and mathematical description of the architecture of the technological systems, methodology of the mathematical description of dynamics of power systems, the method of the graph of cause-effect relations are proposed. The mathematical substantiation is based on a systematic approach to complex mathematical and logical modeling in the technological system as a dynamic system. Decision-making to support the functioning of energy systems is based on a mathematical description of the architecture of technological systems, the methodology of mathematical description of the dynamics of energy systems, the method of the graph of causation. Coordination of energy production and consumption is based on forecasting changes in the parameters of technological processes. A real energy system is a dynamic system, the mathematical model of which reflects the properties of the transformation of influences, that is, its dynamic properties. The maintenance of the functioning of energy systems takes place in the composition of such technological systems, the basis of which is the dynamic system. When designing a technological system, we lay its foundation – an integrated dynamic subsystem for evaluating changes in both production and energy consumption. Based on the system-structural and mathematical substantiation of the architecture of the technological system, we consider the relation category as organizing interactions not only within the elements of the technological system, but also within the elements of the integrated dynamic subsystem, which makes it possible to perform workability control and identify the state of the energy system based on the developed method of the graph of cause and effect relations. The final assessment, obtained at the expense of logical relations in the dynamic subsystem, allows obtaining new properties of the energy system as a result of appropriate decision-making and identification of new operating conditions. Moreover, the relations between the dynamic subsystem and the blocks in its composition allow, based on the assessment of the state of the parameters diagnosed in these blocks, to confirm the new conditions of the energy system functioning.

Chapter 2 – The integrated Smart Grid system of harmonization of production and consumption of electric power and heat in the biogas cogeneration system with the use of heat-pumping power supply of the biogas plant, which uses fermented wort as a low-potential source of power, is proposed. A change in the power factor of the cogeneration system, the temperature of local water is predicted by measuring the voltage at the inlet to the inverter, at the outlet from the inverter and voltage frequency. In the engine cooling circuit, the temperature of cooling water at the inlet to the heat exchanger, at the outlet from the heat exchanger, and the return water temperature are measured. It was proposed to estimate a change in the ratio of voltage at the inlet to the inverter and at the outlet from the inverter. Making forestalling decisions to change the power of the heat pump and the number of plates in the heat exchanger of the engine cooling circuit makes it possible to maintain the voltage at the entrance to the inverter and the temperature of the heated local water. The complex mathematical and logical modeling of the cogeneration system, based on the mathematical substantiation of the architecture of the cogeneration system and mathematical substantiation of the maintenance of functioning of the cogeneration system, was performed. Time constants and coefficients of the mathematical models of dynamics regarding the estimation of a change in the power factor of the cogeneration system, temperature of local water, were determined. Functional estimation of a change in power factor of the cogeneration system in the range of 85–95%, temperature of local water in the range of 30–55°C at the compensation of reactive power of up to 40% was obtained. Determining final functional information provides an opportunity to make forestalling decisions on a change in the power of a heat pump and a change in the number of plates in the heat exchanger of the engine cooling circuit to maintain the functioning of the cogeneration system.

Chapter 3 – The integrated Smart Grid System of harmonization of production and consumption of electric power and heat in the biodiesel cogeneration system is proposed. The integrated dynamic subsystem of the biodiesel cogeneration system includes the following components: the electric network, the cogeneration unit, the engine cooling circuit as a part of the cogeneration plant, the biodiesel plant, the heat exchanger for heating oil with biodiesel heat. A change in the ratio of production and consumption of electric power and heat, change in the temperature of oil and change in the temperature of local water of the engine cooling circuit is predicted by measuring temperature of biodiesel at the outlet from the heat exchanger, the temperature of cooling water at the outlet of the heat exchanger the engine cooling circuit and the temperature of return water Making forestalling decisions to support the supply of fresh oil for heating, to support the supply of heated oil for etherification, changing number of plates of the biodiesel heat exchanger and the changing number of plates of the heat exchanger of the engine cooling circuit makes it possible to maintain the ratio of production and consumption of electric power and heat and the temperature of the heated oil and local water. *The complex mathematical and logical modeling of the cogeneration system, based on the mathematical substantiation of the architecture of the cogeneration system and mathematical substantiation of the maintenance of functioning of the cogeneration system, was performed. Time constants and coefficients of the mathematical models of dynamics regarding the estimation of a change in the temperature of oil, temperature of local water, were determined. Functional estimation of a change in the ratio of production and consumption of electric power and heat of the cogeneration system in the range of 0.6606 – 0.6611, temperature of oil in the range of 20–45°C,* temperature of local water in the range of 30–55°C was obtained. Determining final functional information provides an opportunity to make forestalling decisions to changing the ratio of production and consumption of electric power and heat to maintain the functioning of the cogeneration system.

Chapter 4 – The integrated Smart Grid System of harmonization of production and consumption of electric power and heat in the pellet cogeneration system is proposed. The integrated dynamic subsystem includes the following components: the electric network, the cogeneration plant, the second circuit as a part of the cogeneration plant, the drying chamber, the heat exchanger for heating air with local water of second circuit of the cogeneration plant, air fan. A change in the ratio of production and consumption of electric power and heat, change in the moisture content of the air in the drying chamber and change in the temperature of local water of the second circuit of the cogeneration system is predicted by measuring temperature of air at the outlet from the heat exchanger, the temperature of the of flue gases at the inlet of the heat exchanger and the return water temperature. Making forestalling decisions to support the supply of raw timber for drying, dried timber for production of the pellet fuel, changing number of the number of revolutions of the electric motor of the air fan and the changing number of plates of the heat exchanger of the second circuit makes it possible to maintain the ratio of production and consumption of electric power and heat and the moisture content of the air and local water The complex mathematical and logical modeling of the cogeneration system, based on the mathematical substantiation of the architecture of the cogeneration system and mathematical substantiation of the maintenance of functioning of the cogeneration system, was performed. Time constants and coefficients of the mathematical models of dynamics regarding the estimation of a change in the moisture content of the air, temperature of local water, were determined. Functional estimation of a change in the ratio of production and consumption of electric power and heat of the cogeneration system in the range of 0.6294–0.6305, moisture content of the air in the range of 12–40%, temperature of local water in the range of 90–79.83°C was obtained. Determining final functional information provides an opportunity to make forestalling decisions to changing the ratio of production and consumption of electric power and heat to maintain the functioning of the cogeneration system.

Chapter 5 – The integrated Smart Grid System of harmonization of production and consumption of electric power of the heat pump power supply and hot water power supply in the wind-solar electric system with hybrid solar collectors are developed. The integrated dynamic subsystem of the wind-solar electric system includes the following components: the electric network, wind turbine, photovoltaic solar panels, hybrid solar collectors, grid inverter, heat pump, two-section storage tank, upper section – for hot water supply, lower section – a low-grade energy source, frequency converter. Integrated systems based on predicting changes in the power factor, temperature of local water when measuring voltage from hybrid solar collectors at the input to the grid inverter, voltage at the output of the frequency converter and voltage frequency. The adoption of advanced decisions to maintain the temperature of local water by changing the power of the electric motor of the heat pump compressors and electric motor of the circulation pump based on establishing the ratio of the voltage at the input to the grid inverter and the voltage at the output of the frequency converter are measured. The power factor of the wind-solar electric system is maintained. The complex mathematical and logical modeling of the wind-solar electric system, based on the mathematical substantiation of the architecture of the wind-solar electric system and mathematical substantiation of the maintenance of functioning of the wind-solar electric system, is performed. Time constants and coefficients of the mathematical models of dynamics regarding the estimation of a change in the power factor of the system, the temperature of local water is predicted when measuring voltage from hybrid solar collectors at the input to the grid inverter, voltage at the output of the frequency converter and voltage frequency. Functional estimation of a change in power factor of the wind-solar electric system in the range of 58–98%, temperature of local water in the range of 30–55°C for hot water power supply and temperature of local water in the range of 35–55°C for heat pump power supply. Determining final functional information provides an opportunity to make forestalling decisions on a change in the of the electric motor of the heat pump compressors and electric motor of the circulation pump to prevent the peak load of the power system under voltage regulation conditions when connecting the heat pump power supply and hot water supply.

Chapter 6 – The integrated Smart Grid System of harmonization of production and consumption of electric power based on a prediction of changes in the battery capacity in photoelectric charging station is developed. The integrated dynamic subsystem of the photoelectric charging station includes the following components: mains, photoelectric solar panels, a hybrid inverter, rechargeable batteries, a two-way Smart Meter counter and a charger. Advanced decisions on the change in power transmission capacity have made it possible to regulate voltage in the distribution system by maintaining the power factor of the photoelectric charging station. Voltages at the input to the hybrid inverter and in the distribution system were measured to assess their ratio. Comprehensive mathematical and logical modeling of the photoelectric charging station was performed based on the mathematical substantiation of architecture and operation maintenance. A dynamic subsystem including such components as mains, a photoelectric module, a hybrid inverter, batteries, a two-way counter Smart Meter and a charger formed the basis of the proposed technological system. Time constants and coefficients of mathematical models of dynamics in terms of estimation of changes in the battery capacity and power factor of the photoelectric charging station were determined. A functional estimate of changes in the battery capacity and power factor of the photoelectric charging station was obtained. Maintenance of voltage in the distribution system was realized based on resulting operation data to estimate a change in the battery capacity. Advanced decision-making has made it possible to raise the power factor of the photoelectric charging station up to 40% due to matching the electric power production and consumption.

Maintenance of operation of the photoelectric charging station using the developed Smart Grid technology has enabled prevention of peak loading of the power system due to a 20% reduction of power consumption from the network.

Chapter 1

Methodological and Mathematical Substantiation of Smart Grid Technologies for Maintaining the Functioning of Electric Systems

Abstract

The mathematical substantiation of maintenance of the operation of the technological system Smart Grid based on the architecture and mathematical description of the architecture of the technological system, methodology of the mathematical description of dynamics of power systems, the method of the graph of cause-effect relations is proposed. The mathematical substantiation is based on a systematic approach to complex mathematical and logical modeling in the technological system as a dynamic system. Decision-making to support the functioning of energy systems is based on a mathematical description of the architecture of technological systems, the methodology of mathematical description of the dynamics of energy systems, the method of the graph of causation. Coordination of energy production and consumption is based on forecasting changes in the parameters of technological processes.

A real energy system is a dynamic system, the mathematical model of which reflects the properties of the transformation of influences, that is, its dynamic properties. The maintenance of the functioning of energy systems takes place in the composition of such technological systems, the basis of which is the dynamic system. When designing a technological system, we lay its foundation — an integrated dynamic subsystem for evaluating changes in both production and energy consumption Based on the system-structural and mathematical substantiation of the architecture of the technological system, we consider the relation category as organizing interactions not only within the elements of the technological system, but also within the elements of the integrated dynamic subsystem, which makes it possible to perform workability control and identify the state of the energy system based on the developed method of the graph of cause and effect relations. The final assessment, obtained at the expense of logical relations in the dynamic subsystem, allows obtaining new properties of the energy system as a result of appropriate decision-making and identification of new operating conditions. Moreover, the relations between the dynamic subsystem and the blocks in its composition allow, based on the assessment of the state of the parameters diagnosed in these blocks, to confirm the new conditions of the energy system functioning.

Keywords: Smart Grid technologies, forecasting changes in process parameters, harmonization of energy production and consumption

Introduction

Distributed generation of electrical energy using renewable sources is due to daily and seasonal fluctuations in energy production with uneven consumption. The process of transition to alternative energy - biofuel, solar, wind, with the stochastic nature of energy production, requires intelligent systems for managing the flow of electric energy and consumption Smart Grid technologies, demand management systems and energy storage are new components for the integration of distributed energy generation in the energy system An urgent further development in this direction is predicting changes in technological process parameters to coordinate production and energy consumption. The presented materials as a result of the research are based on a systematic approach. Mathematical description of architecture of electric systems, methodology of mathematical description of dynamics of power systems, method of causal graphs, mathematical description of Smart Grid of support of functioning of electric systems are offered and tested. Integrated Smart Grid systems for coordination of production and consumption of the electric power on the basis of forecasting of change of parameters of technological processes are developed. The presented materials have been tested at international scientific conferences, in the scientific journals, including referenced in the database Scopus (9 personal publications), when teaching disciplines at the Odessa Polytechnic National University: "A systematic approach in the study of non-traditional energy objects"; "Wind energy installations"; "Humidification and diagnostics of energy equip-ment"; "Mathematical problems of renewable energy"; "Production and use of biofuel"; "Energy technological installations"; "Systems and modes of energy supply"; "Computer technology and algorithmic languages." Under the scientific supervision of the author, bachelor's and master's works of students, dissertations of three graduate students were prepared. One of the dissertations was defended at the Dissertation Defense Council. The applicant was awarded the degree of candidate of technical sciences. The author published 245 printed scientific works in national and international databases.

Methodological and Mathematical Substantiation of the Architecture of Electric Systems

One of the main properties of energy systems is the mandatory exchange of substance, energy and information with the environment. The functioning of energy systems can be considered, in this regard, as a reproduction of external and internal influences and change in initial conditions. The nature of the reaction is determined by the inertia of the devices and the speed of transient processes, that is, by the dynamic properties that appear during the functioning of energy systems, information about which can be obtained from the dynamic characteristics. The dynamic characteristics of energy systems can be described by a finite set of parameters with respect to changes in time and by a spatial coordinate that coincides with the direction of the medium's flow. Input influences can also be described by a set of parameters. Therefore, the dynamic description of the energy system most fully and multifacetedly characterizes its functioning. Thus, we determine that a real energy system is a dynamic system, the mathematical model of which reflects the properties of the transformation of influences, that is, its dynamic properties. The maintenance of the functioning of energy systems takes place in the composition of such technological systems, the basis of which is the dynamic system. When designing a technological system, we lay its foundation – an integrated dynamic subsystem for evaluating changes in both production and energy consumption (Figure 1.1).

Representing the construction of a technological system as an organization of a complex system, we expand it by building around its base - an integrated dynamic subsystem other modules predicting the basic components of the technological process (Figure 1.1). The relationship of the dynamic subsystem with other modules of the technological system, which assess the support of the technological process, prevention of its violation and functional efficiency in decision-making conditions, is based on mathematical modeling of their logical connections that change over time. This provides an opportunity to set new properties of the integrated dynamic subsystem – the energy system and modules of the technological system. Mathematical understanding of the principle of system organization is based on the application of logical structures that describe "organizing" relationships between system elements. This provides an opportunity to evaluate the organization's processes using system-structural analysis. Based on this provision, we consider the category of relations as cause-and-effect relationships that lead to the coordinated behavior of the dynamic subsystem and other elements of the technological system.

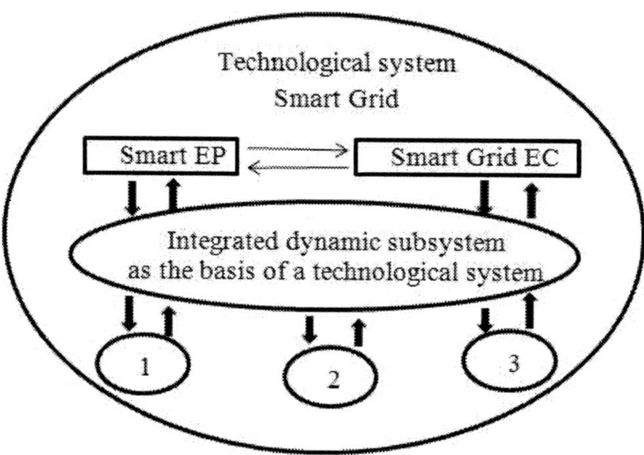

Figure 1.1. Architecture of the technological system: EP – energy production; EC – energy consumption; 1 – units of technological process support; 2 – units for changing operational conditions; 3 – units of estimation of functional efficiency.

The methodological justification of the proposed architecture of the technological system can be presented as follows:

{(technological system), (integrated dynamic subsystem) →
dynamic properties → logical relations → properties
(state identification) → (integrated dynamic subsystem) →
logical relations → properties (modules of technological systems) →
(technological system)}.

The mathematical substantiation of the architecture of the technological system (1) is presented: (Figure 1.2).

$$TS(\tau)=\left\{\begin{matrix}\left[\begin{matrix}ID(\tau)(P(\tau)\left\langle\begin{matrix}x_0(\tau),x_1(\tau),x_2(\tau),f(\tau),K(\tau),\\ y(\tau,z),d(\tau),FI(\tau)\end{matrix}\right\rangle,LC(\tau),\\ \langle f(\tau),K(\tau),y(\tau,z),d(\tau),FI(\tau)\rangle LS(\tau),P(\tau)),\end{matrix}\right]\\ R(\tau),(P_i(\tau)\langle x_1(\tau),f_i(\tau),K_i(\tau),y_i(\tau),FI(\tau)\rangle LS(\tau))\end{matrix}\right\} \quad (1)$$

Figure 1.2. Mathematical substantiation of the architecture of the technological system.

where $TS(\tau)$ – technological system; τ – time, s; $ID(\tau)$ – integrated dynamic subsystem $P(\tau)$ – properties of the components of the technological system; $x(\tau)$ – impacts; $f(\tau)$ – parameters that are measured $K(\tau)$ – coefficients of the mathematical description of the dynamics of changes in the predicted parameters; $y(\tau, z)$ – predicted output parameters; z – length coordinate coinciding with the direction of the flow of the medium, m; $d(\tau)$ – dynamic parameters of predicted parameter changes;; $FI(\tau)$ – functional resulting information; $LC(\tau)$ – logical relations regarding the control of the operational efficiency of the technological system; $LS(\tau)$ – logical relations regarding the identification of the state of the technological system; $R(\tau)$ – logical relations in $TS(\tau)$ to confirm the correctness of the decisions taken by the blocks in the technological system. Indices: i – the number of technological system elements; 0, 1, 2 – initial stationary mode, external and internal nature of influences.

The Methodology of Mathematical Description of the Dynamics of Electric Systems

Forecasting the parameters of the technological process to support the functioning of the technological system at the decision-making level requires an important condition. This is when the movement of the system occurs in the non-linear region of its space. Irreversible thermodynamic processes taking place in a dynamic energy system should be represented by nonequilibrium thermodynamic equations, where state parameters are considered as continuous functions of spatial coordinates and time. The nonlinearity of the dynamic processes occurring in the technological system determines the mathematical modeling of the dynamics relative to the significant parameters that are predicted. These are parameters selected on the basis of expert knowledge, the properties of which are decisive for the functioning of this energy system. The system of differential equations includes the equation of state as an estimate of the physical model of the system, the energy equation of the transmitting and receiving media, and the heat balance equation for the heat exchanger wall. If necessary, the system of differential equations can be supplemented with a continuity equation, if the change in the flow rate of the working environment is chosen as an essential parameter to be predicted. The development of a mathematical model begins with an assessment of the physical model of the system expressed by the equation of state. Further, developing a mathematical model, the equations of state are supplemented with energy equations of the transmitting and receiving media, which are considered as equations of enthalpy transfer. Equations of the energy of the receiving environment, characterized by the presence of a significant predicted parameter, must be drawn up with a representation of the change of these parameters not only in time, but also along the spatial coordinate, which coincides with the direction of the medium flow.

A distinctive feature of mathematical models is precisely these energy equations, since they contain terms that reflect a wide range of influences coming to the input of a functioning energy system from the environment. There is a need to obtain an estimate of the change of a significant parameter, which is predicted, based on a wide range of disturbances coming from the environment. In this case, the result of the solution of the system of differential equations is the transfer functions obtained on the basis of the Laplace transform, which evaluate the change of the essential parameters, which are predicted, when the parameters of the technological process change. Analyzing the obtained mathematical model, we set the internal parameters that are part of the coefficients of the dynamics equations. The linearization performed for the solution of the system of nonlinear differential equations is correct. In the real conditions of the functioning of the energy system, during the transition from stationary states and the arrival of external and internal influences, the coefficients of the dynamics equations are rearranged in time due to the change of the internal parameters that are measured. Thus, according to the mathematical justification of the architecture of technological systems, the change in the properties of the dynamic subsystem is due to the change in the initial conditions of operation. This is the reason for obtaining an estimate of the state of the internal parameter to be measured, the coefficient of the transfer function, the significant parameter to be predicted, the dynamic parameters of the dynamic characteristics of the significant parameter to be predicted. The final assessment, obtained at the expense of logical relations in the dynamic subsystem, allows obtaining new properties of the energy system as a result of appropriate decision-making and identification of new operating conditions. Moreover, the relations between the dynamic subsystem and the blocks in its composition allow, based on the assessment of the state of the parameters diagnosed in these blocks, to confirm the new conditions of the energy system functioning.

Cause-and-Effect Graph Method

Based on the system-structural and mathematical substantiation of the architecture of the technological system (1), we consider the relation category as organizing interactions not only within the elements of the technological system, but also within the elements of the integrated dynamic subsystem, which makes it possible to perform workability control control and identify the state of the energy system based on the developed method of the graph of cause and effect relations (Figure 1.3).

Thus, block CT_1 evaluates the change in the initial conditions of functioning due to the appearance of influences. Next, using a chain of cause-and-effect relationships, when the previous evaluation of an event that occurs is the reason for the acquisition of the next one, we receive summary information from the CTc control unit. Adopting anticipatory decisions provides an opportunity to maintain the technological process or to change the regime conditions of operation in order to prevent disruption of the technological process. Moreover, there is an opportunity to functionally evaluate the change in the efficiency of the technological process and the efficiency of decision-making. After the appropriate decision-making, the new conditions of the energy system functioning are confirmed using the second part of the graph of cause-and-effect relationships with respect to the predicted parameters, estimated according to the first part of the graph. Making anticipatory decisions and checking the correctness of decision-making, which take place on the basis of logical connections in the dynamic subsystem, are final if the dynamic subsystem receives confirmatory evaluations of the correctness of decision-making from the corresponding blocks as part of the technological system.

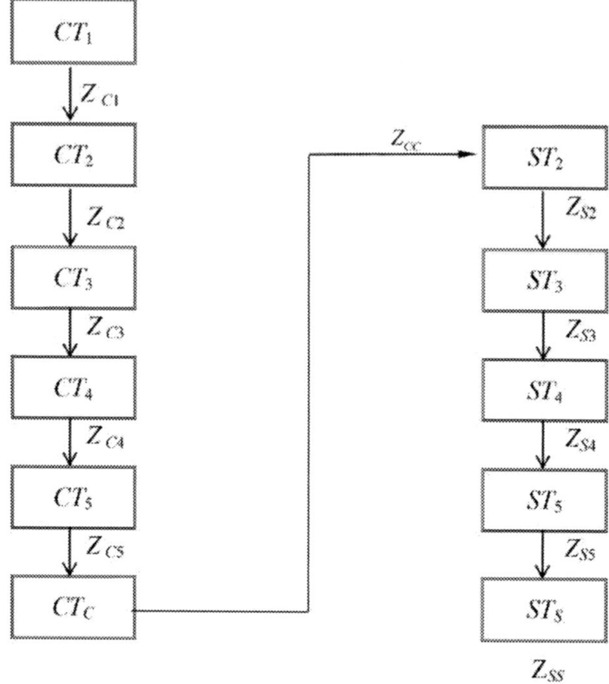

Figure 1.3. Graph of cause-and-effect relationships of the integrated dynamic subsystem: *CT* – event control; *Z* – logical relations; *ST* – event identification. Indexes: 1 – influences; 2 – internal parameters that are measured; 3 – coefficients of dynamics equations; 4 – essential parameters that are predicted; 5 – dynamic parameters; *c* – workability control; *s* – state.

Mathematical and Logical Substantiation of the Smart Grid Technologies to Maintain the Operation of Electric Systems

The mathematical substantiation of maintenance of the operation of the technological system Smart Grid (2), (Figure 1.4), based on the mathematical description of the architecture of the technological system (1) (Figure 1.2.), methodology of the mathematical description of dynamics of power systems, the method of the graph of cause-effect relations (Figure 1.3) is proposed.

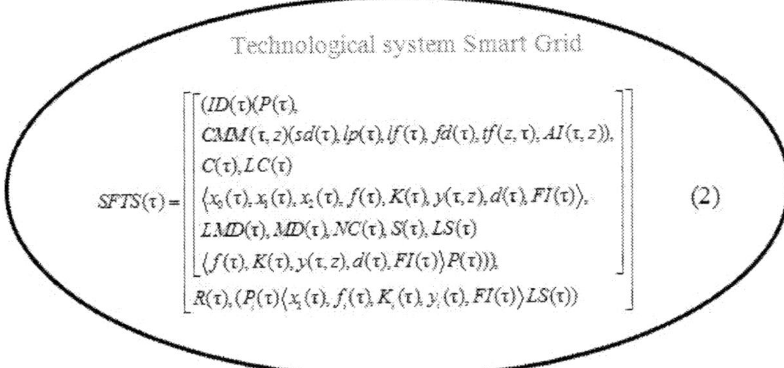

Figure 1.4. Mathematical substantiation for supporting the functioning of the technological system.

where $SFTS(\tau)$ – Smart Grid maintenance of the operation of the technological system; τ – time, s; $ID(\tau)$ – integrated dynamic subsystem; $P(\tau)$ – the properties of the elements of the integrated dynamic subsystem, units of the technological system; $CMM(\tau, z)$ – complex mathematical modeling of the dynamics of changes predicted parameters; $sd(\tau)$ – the input data; $lp(\tau)$ – the boundary change in parameters; $lf(\tau)$ – the levels of operation; fd – the obtained parameters; $tf(\tau,z)$ – the transfer function of predicted parameters; $AI(\tau,z)$ – the standard information regarding the evaluation of the maximum admissible change predicted parameters; $C(\tau)$ – the control of workability of the technological system; $LC(\tau)$ – the logical relations of the control of the technological system; $x(\tau)$ – impacts; $f(\tau)$ – the measured parameters; $K(\tau)$ – the coefficients of the mathematical description of the dynamics of changes predicted parameters; $y(\tau, z)$ – the output predicted parameters; z – length coordinate coinciding with the direction of the flow of the medium, m; $d(\tau)$ – the dynamic parameters of estimation of a change in the predicted parameters; $FI(\tau)$ – functional resulting information on decision making; $LMD(\tau)$ – the logical relations of decision making; $MD(\tau)$ – decision making; $NC(\tau)$ – the new conditions of the technological system; $S(\tau)$ – the identification of the state of the technological system; $LS(\tau)$ – the logical relations of identification of the state of the technological system; $R(\tau)$ – the logical relations between the dynamic subsystem and units as part of the technological system/Indices: i – the number of elements of $SFTS(\tau)$; 0, 1, 2 – the initial, external, and internal character of influences.

Mathematical substantiation of the architecture of the technological system (1) (Figure 1.2) and mathematical substantiation of maintenance of the operation of the technological system Smart Grid (2) (Figures 1.3, 1.4) make it possible to maintain the operation of the technological system using the following actions:

- Workability control ($C(\tau)$) of the dynamic subsystem based on complex mathematical ($CMM(\tau, z)$) and logical ($LC(\tau)$) modeling regarding obtaining standard ($AI(\tau,z)$) estimate of a change in the predicted parameters;
- Workability control ($C(\tau)$) of the dynamic system based on complex mathematical ($CMM(\tau, z)$) and logical ($LC(\tau)$) modeling regarding the obtaining functional ($FI\ (\tau)$) estimate of a change in the predicted parameters;
- Making anticipatory decisions to support the functioning of the technological system ($MD(\tau)$) with the use of the functional resulting information ($FI\ (\tau)$), obtained based on logical modeling ($LMD(\tau)$);
- Identification ($S(\tau)$) of the new conditions of functioning of the technological system ($NC(\tau)$) based on logical modeling ($LS(\tau)$) as a part of the dynamic subsystem and confirmation of new operating conditions based on logical modeling ($R(\tau)$) from the units of the technological system.

Obtaining the resulting information on the basis of energy system performance control allows you to make anticipatory decisions on the establishment of new operating conditions, confirmed on the basis of identification information in the dynamic subsystem and blocks in the technological system.

Results and Discussion

An example would be using methodological and mathematical substantiation (1), (2), (Figures 1.1–1.4) the author, Chaikovskaya, E. (2017) proposes an energy-saving technology to support changes in the battery capacity. The basis for coordinating the processes of heat and mass transfer during charging and discharging the battery is to predict the change in the charge and discharge voltage when measuring the temperature of the electrolyte at the inlet and outlet of the battery. For this purpose, a mathematical model of the dynamics of changes in the battery voltage during charging and discharging has been developed. The transfer functions for the channels "charge voltage – electrolyte temperature in the battery volume," "discharge voltage – electrolyte temperature in the battery volume" make it possible to predict the change in the charge and discharge voltage based on an assessment of the electrolyte temperature change in the pores of the plates and above the plates. Transfer functions are obtained as a result of solving a system of nonlinear differential equations using the Laplace transform method.

The system of differential equations includes the equation of state as an estimate of the physical model of the battery, the equations of charge and discharge energy, the heat balance equation for the wall of the battery plates. The charge and discharge energy equations were developed using an estimate of the change in the electrolyte temperature in the pores of the plates and above the plates both in time and along the spatial coordinate of the battery plates. The integrated voltage change system, obtained on the basis of the coordination of heat and mass transfer processes during discharge and charge, makes it possible to make timely decisions on recharging the battery in order to prevent overcharging and unacceptable discharge. The electrochemical and diffusion processes were aligned, which accompany charging and discharging the battery, to make preemptive decisions to support a change in the capacity of accumulator battery and to prevent gas formation.

An analytical estimation of the change in voltage and discharge voltage without connecting the load and when connecting the load was obtained. The functional system of change in the voltage of accumulator battery was developed, which makes it possible to maintain capacity of the battery based on the prediction of change in the discharge voltage when measuring the temperature of electrolyte in the volume of accumulators using switching to charge. This makes it possible to provide maintaining the capacity and to prevent the recharge and unacceptable battery discharge. Determining the exact time of charge before the start of gas formation makes it possible to reduce the charge time to save electricity and to prevent the formation of gas. In the conditions of operation of a wind power plant with a power of, for example, 10 kW, shortening the charge period and preventing gas formation reduces the cost of energy production and the payback period of the wind power plant by up to 25%.

To give another example, using methodological and mathematical substantiation (1), (2), (Figures 1.1–1.4) the author, Chaikovskaya, E. (2019) proposes an energy-saving technology to support changes an energy-saving technology to support the functioning of the wind-solar electrical system. An integrated system to maintain functioning of a wind-solar electric system has been built, based on adjusting the generation and consumption of energy in terms of energy saving. Typically, a hybrid charge controller included in a wind-solar electric system maintains the charge of a rechargeable battery by using a thermoelectric battery as the non-regulated ballast. Resetting excess energy to the ballast when using the MPPT function of the controller leads to the uncompensated-for losses of electrical energy, which does not make it possible to ensure an appropriate level of capacity for the charge of a rechargeable battery.

Moreover, the use of a thermoelectric battery as a ballast eliminates the need to maintain the operation of a wind energy installation by using the accumulation of heat in order to regulate the power of a wind turbine. Failure to account for this property of a thermoelectric battery could lead to the acceleration of the wind turbine at considerable wind speed and to its malfunctioning. It is known that a thermoelectric battery is controlled in line with the thermostat principle, that is, when establishing the required temperature of heated local water, the thermoelectric battery is disconnected from power. Not using a change in the local water flow rate during the period of thermoelectric battery charge, when changing capacity, prolongs the duration of charge and leads to considerable expenditures on electricity.

A technique to overcome these difficulties has been proposed. It is the thermoelectric battery that must become the main center of adjusting a change in the total power of a wind-solar electric system to power consumption by redistributing the accumulated heat and electric energy in terms of consumption. It has been proposed to predict a change in the capacity of a rechargeable battery when measuring the total voltage at the input to a hybrid charge controller and voltage at the output from the inverter to estimate the ratio of electrical energy generation and consumption when measuring the voltage frequency. For this purpose, as a result of solving a system of non-linear differential equations, transfer functions were obtained for the channels "accumulator battery capacity – power of the thermal accumulator," "circulation pump electric motor speed – voltage frequency," "local water consumption – circulation pump electric motor speed."

The system of differential equations includes the equation of state as an estimate of the physical model of the electrical system, the energy equation of the transmitting and receiving media – the heater of the thermal accumulator and local water, respectively, the heat balance equation for the wall of the thermal electric heater. To estimate the change in the flow of local water, the system of differential equations is supplemented with a continuity equation. The equation for the energy of the receptive medium was developed on the basis of an estimate of the change in the temperature of the local water both in time and along the spatial coordinate coinciding with the direction of the flow of the medium. Making preliminary decisions on changing the power of a thermoelectric battery makes it possible, while maintaining the capacity of the rechargeable battery, to ensure a change in the temperature of heated local water, based on changing the number of rotations of the circulating pump's electric motor in terms of water flow rate, thereby reducing the charge duration by up to 30%.

Chapter 2

Smart Grid Technology for Maintaining the Functioning of a Biogas Cogeneration System

Abstract

The integrated Smart Grid System of harmonization of production and consumption of electric power and heat with the use of heat-pumping power supply of the biogas plant, which uses fermented wort as a low-potential source of power, was developed. A change in the power factor of the cogeneration system, the temperature of local water is predicted by measuring the voltage at the inlet to the inverter, at the outlet from the inverter and voltage frequency. In the engine cooling circuit, the temperature of cooling water at the inlet to the heat exchanger, at the outlet from the heat exchanger, and the return water temperature are measured. It was proposed to estimate a change in the ratio of voltage at the inlet to the inverter and at the outlet from the inverter.

Making forestalling decisions to change the power of the heat pump and the number of plates in the heat exchanger of the engine cooling circuit makes it possible to maintain the voltage at the entrance to the inverter and the temperature of the heated local water. The complex mathematical and logical modeling of the cogeneration system, based on the mathematical substantiation of the architecture of the cogeneration system and mathematical substantiation of the maintenance of functioning of the cogeneration system, was performed.

Time constants and coefficients of the mathematical models of dynamics regarding the estimation of a change in the power factor of the cogeneration system, temperature of local water, were determined. Functional estimation of a change in power factor of the cogeneration system in the range of 85–95%, temperature of local water in the range of 30–55°C at the compensation of reactive power of up to 40% was obtained. Determining final functional information provides an opportunity to make forestalling decisions on a change in the power of a heat pump and a change in the number of plates in the heat exchanger of the engine cooling circuit to maintain the functioning of the cogeneration system.

Keywords: Smart Grid technologies, power factor, cogeneration system, biogas plant, heat pump, frequency converter

Introduction

Given the need to save natural fuel and reduce the harmful emissions into the atmosphere, the distributed generation of electric power using renewable sources requires Smart Grid technologies for the integrated control of the flows of electric power and consumption (the author's work, 2022). A load of distributed energy generation is known to consume both active and reactive power, which make full capacity and have different cost values. Active power, as the ratio of active power to full power, is evaluated by power factor cosφ, where φ is the angle of phase shift between current and voltage.

Active power is aimed at the performance of useful action, reactive power is a measure of power exchange between the generator and inductive load. Reactive power is directed at creating magnetic fields, without which functioning of the inductive load is impossible. To assess reactive power, one uses tgφ, which is related to active power by the following ratio: $\cos\varphi=1/\sqrt{(1+tg^2\varphi)}$. In the electric network, there should be a balance of generation and consumption of active and reactive power. While the basic indicator of active power maintenance is the frequency in the power system, the indicator of maintenance of reactive power is the voltage in the distribution network. It is a change in reactive power at the consumption level that is a necessary part of voltage maintenance.

The use of special compensating devices at the consumption level, for example, synchronous compensators, intelligent inverters, etc., capable of both generating and absorbing reactive power, is a necessary component in terms of compensation of a voltage change. The cost of the plant and maintenance of these plants can be quite high. A relevant task in the context of further development of distributed electricity generation is to maintain the balance of production and consumption of active and reactive powers of the cogeneration system based on the heat-pumping power supply of the biogas plant, a low potential source of power for which is fermented wort (the author's works, 2015, 2018). Thus, the author's work (2015) represents energy-saving technology of maintenance of the biogas plant operation, that] makes it possible to predict a change in fermentation temperature and establish the temperature of warming a heat carrier at the inlet to the heat exchanger, built-in the methane tank, with the use of heat pumping power supply. The warming heat carrier temperature is measured at the outlet to the heat exchanger during the period of biogas production. The use of an integrated system for the assessment of changes in fermentation temperature, obtained based on mathematical and logical modeling, ensures a constant release of biogas, timely unloading of fermented wort, and loading fresh material while maintaining the balance of the flows of fresh and fermented raw materials. The prediction of a change in the fermentation temperature requires making advanced decisions about a change in the temperature of the warming heat carrier at the inlet to the heat exchanger, fitted in the methane tank.

Thus, the author's work (2018) represents the energy-saving technology of maintaining the operation of heat pump power supply, that provides an opportunity to make advanced decisions to change the number of rotations of the heat pump compressor of the electric engine. A change in the temperature of heated water takes place at the ratio of the established ratio of the cooling agent at the outlet from the heat pump capacitor and evaporation pressure at the outlet of the evaporator.

Maintenance of the power factor of the cogeneration system on biogas fuel should take place at the level of frequency control of the electric engine of the heat pump compressor as a consumer of active and reactive power in ensuring the constant release of biogas as the producer of the alternative power source. Voltage regulation in the distribution system will make it possible to set a new level of active power, both electric and thermal cogeneration systems. Moreover, it is necessary to predict a change in the temperature of local heated water, in the heat exchanger of the engine cooling circuit of the cogeneration system, under conditions of changing the ratio of production of electric power and heat at a change in consumption. For this purpose, it is necessary to predict a change in power factor of the cogeneration system during the measurement of the voltage at the inlet in the inverter and at the outlet from the inverter on terms of the estimation of their ratio and voltage frequency. Making advanced decisions about a change in power of the heat pump energy supply of the biogas plant and a change in the number of plates of the heat exchange of the engine cooling circuit makes it possible to estimate a change in power factor of the cogeneration system. The ratio of production of electric power and heat at a change in consumption is maintained.

Optimization of distributed generation of electric power traditionally uses the improvement of intelligent control systems both by the production of electrical power, and consumption. Thus, the paper (Daniyan, I., Daniyan, O., & Abiona, O., Mpofu, K., 2019) presents the Smart system of biogas production based on a microcontroller for the estimation of parameters of biogas plant operation. Detection of the interaction between a change in the parameters offered an opportunity to conclude the necessity of continuous monitoring of biogas production process in terms of its optimization. The paper (Gholizadeh, T., Mohammad V., & Rostamzadeh, H., 2019) presented the innovative system of electric power cogeneration for cooling to be integrated into the cycle of biogas turbines. The capacity of the implemented integrated power plant was studied using the laws of thermodynamics and software Engineering Equation Solver (EES). The total cooling load and net electrical power were evaluated: 505.2 kW and 1168 kW, respectively. Energy and exergetic efficiency of the proposed innovation system was presented. Thus, thr paper (Rostampour, V., Jaxa-Rozen, M., & Bloemendal, M., Kwakkel, V., Keviczky, T., 2019) presents the results of intelligent control of the distributed power generation based on large-scale seasonal heat accumulation (ATES).

Article (Yinan, L., Wentao, Y., & Ping, H., Chang, Ch., Xiaonan, W., 2019) proposed a hierarchical framework for the control of electric power consumption by one-rank information exchange in the real-time market. The results of the optimization of electricity consumption schedule based on the interaction between electric power generation and participants of consumption are presented. The results of the implementation of the algorithm of stochastic optimization of distributed electricity generation using fuzzy logic are presented in paper (Saad, A., Samy, F., & Osama, M., 2019). The relation of electric system loading and operational costs with the flexibility of control of distributed generation was found. The boundary level of electricity generation using a communal network as virtual storage in terms of maintaining the control flexibility was proposed. Research (Perera, A., Vahid M., & Wickramasinghe, P., Scartezzini, J., 2019) proposed the cyber-physical system of control of distributed generation of electric power, based on the theory of consensual protocol.

It is known that electric devices, for example, asynchronous electric engines, along with active power, consume reactive power as well. Reactive power does not perform a useful function directly but maintains the operation of these devices to create magnetic fields, without which power consumption is impossible. The well-known VVO concept offers the opportunity to change the electrical energy consumption based on the voltage regulation in the distribution system with the use of a change in reactive power. Research (Davye, M., Daranith, & Ch., Dae-Hyun, Ch., 2020) proposed the intelligent transducer regulating voltage by absorbing or supplying reactive power (*Var*) to the network or from the network using the *Volt* – *Var* control function. This article deals with the capacitive (that is, *Var*-injection) and inductive (that is, *Var*-absorption) effects using the intelligent inverter and its ability to affect the voltage at the distribution level.

When the intelligent converter introduces reactive power, it increases the distribution voltage. Conversely, the voltage decreases when the intelligent inverter absorbs reactive power. Paper (Xiqiao, L., Yukun, L., & Xianhong, B., 2019) presents the VVO optimization model to establish priority sensitivity to data changes, based on precise measurement in order to improve the programs of response to electricity consumption. Traditionally, the use of reactive power to control the voltage in an electric network is carried out by means of, for example, transformers, capacitor batteries, voltage regulators, static synchronous compensators, etc. The installation and maintenance costs of these devices can be quite high and some have a relatively slow response time of about a few seconds (Davye, M., Daranith, & Ch., Dae-Hyun, Ch., 2020). Heat pump power supply of the biogas plant provides an opportunity to use a heat pump as a voltage regulator in the distribution network. A change in the reactive power of the cogeneration system is estimated based on the evaluation of a change in the number of revolutions of the electric engine of the heat pump compressor within the term of biogas production.

That is why it is proposed to measure voltage at the inlet to the inverter and voltage at the outlet from the inverter in terms of estimation of their ratio when measuring voltage frequency. Moreover, the use of heat pumping power supply of the biogas plant makes it possible to regulate the ratio of production of electricity and heat at changing consumption. That is why it is proposed to measure the temperature of cooling water at the inlet to the heat exchanger of the engine cooling circuit and at the outlet from the heat exchanger and the temperature of cooling water. The engine cooling circuit, which usually acts as a protective element for the engine of the cogeneration system to prevent overheating, becomes a complex information system to assess changes in an electrical and thermal power cogeneration system.

It is this energy equipment that responds to a change in consumption of electric power regarding the frequency control of the heat pumping power supply of the biogas plant in regard to the change in the balance of electrical power and heat consumption. That is why for maintaining the operation of the cogeneration biogas fuel system it is necessary to predict of a change in the power factor of the cogeneration system and the temperature of local heated water. Making advanced decisions on changing the power of a heat pump and changing the number of the plates of the heat exchanger of the engine cooling circuit makes it possible to regulate voltage in the distribution system.

Ensuring a change in active electric and thermal powers of the cogeneration system makes it possible to maintain their ratio at a change in consumption.

Methodological and Mathematical Substantiation

Based on the methodological, mathematical, logical substantiation of the technological systems (Chapter 1), energy-saving technologies of biogas production with the use of heat-pumping energy supply (the author's works, 2015, 2018) the architecture, mathematical substantiation of the architecture (1), mathematical substantiation of maintenance of the operation (2) of the biogas cogeneration system Smart Grid are proposed (Figure 2.1).

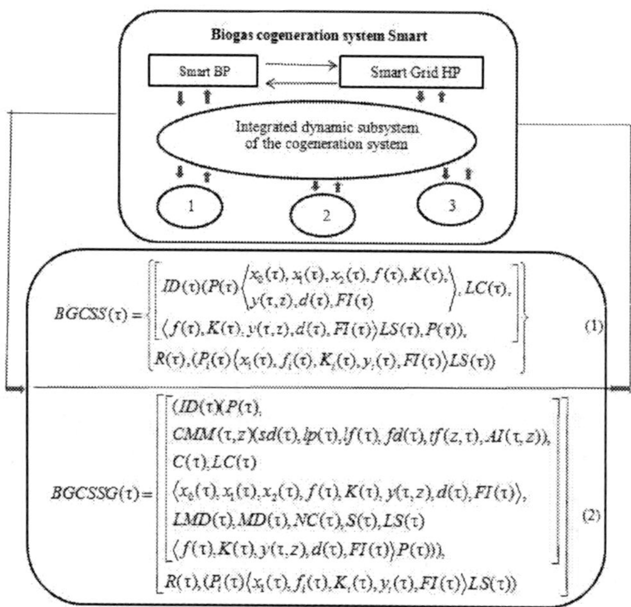

Figure 2.1. Biogas cogeneration system, Smart Grid. The architecture: BP – biogas plant; HP – heat pump system; 1– charge unit; 2–discharge unit; 3 – unit of evaluation of functional efficiency. Mathematical substantiation of the architecture (1). Mathematical substantiation of maintenance of the operation (2).

A cogeneration system is a dynamic system, the operation of which is the reproduction of a change in external, internal influences and initial conditions, for example, a change in power of the electric engine of the heat pump, a change in the ratio of production of electric power and heat under conditions of a consumption change, etc. That is why, when designing a cogeneration biogas system, underlying which is an integrated dynamic subsystem (Figure

2.1). The integrated dynamic subsystem includes the following components: the electric network, the cogeneration unit, the engine cooling circuit as a part of the cogeneration plant, the biogas plant, the heat pump, the frequency converter. The heat pump uses fermented wort as a low-potential power source to maintain the fermentation temperature, unloading of fermented wort, and loading of fresh material. The frequency converter performs the frequency regulation of the electric motor of the spiral compressor of the heat pump. Other units included in the cogeneration system are the units of charge, discharge, and estimation of functional efficiency, which are in harmonized interaction with the integrated dynamic subsystem (Figure 2.1).

The mathematical substantiation of the architecture of the biogas cogeneration system Smart (1), (Figure 2.1), based on the methodology of the mathematical description of dynamics of power systems, the method of the graph of cause-effect relations [Chapter 1] is proposed (Figure 2. 1).

Where $BGCSS(\tau)$ is the biogas cogeneration system Smart; τ is the time, s; $ID(\tau)$ is the integrated dynamic subsystem (the electric network, the cogeneration unit, the engine cooling circuit as a part of the cogeneration plant, the biogas plant, the heat pump, the frequency converter); $P(\tau)$ is the properties of the elements of the cogeneration system; $x(\tau)$ is the influences (a change in parameters: the temperature of heating water at the outlet from the heat exchanger fitted in the methane tank, the temperature of refrigerant at the outlet from the condenser, evaporation pressure, condensation pressure, voltage at the inlet to the inverter and at the outlet from the inverter, voltage frequency, temperature of cooling water at the inlet to the heat exchanger, at the outlet from the heat exchanger, the temperature of return water; $f(\tau)$ is the measured parameters (the temperature of heating water at the outlet from the heat exchanger fitted in the methane tank, the temperature of refrigerant at the outlet from the condenser, evaporation pressure, voltage at the inlet of the inverter and at the outlet of the inverter, voltage frequency, the temperature of cooling water at the inlet of the heat exchanger, at the outlet of the heat exchanger, the temperature of return water); $K(\tau)$ is the coefficients of the mathematic description of the dynamics of a change in the temperature of fermentation, the consumption of refrigerant vapor, the number of revolutions of the compressor electric motor, the temperature of heating water at then outlet from the condenser, power factor, the temperature of local water; $y(\tau, z)$ is the output parameters (the temperature of fermentation, the consumption of refrigerant vapor, the number of revolutions of the compressor electric motor, the temperature of heating water at the outlet from the condenser, power factor, reactive power factor, active, reactive powers of the cogeneration system, temperature of local water); z is the coordinate of the length: the heat exchanger fitted in the methane tank, heat pump condenser, of the heat exchanger of the engine cooling circuit, m; $d(\tau)$ is the dynamic parameters of a change in the temperature of fermentation, the consumption of refrigerant vapor, the number of revolutions of the compressor electric motor, the temperature of heating water at the outlet from the heat pump condenser, the power factor of the cogeneration system, the temperature of local water; $FI(\tau)$ is the functional final information; $LC(\tau)$ is the logical relations regarding the control of the cogeneration system workability; $LS(\tau)$ is the logical relations regarding the identification of the state of the cogeneration system; $R(\tau)$ is the logical relations in $BGCSS(\tau)$ to confirm the correctness of decisions made from the units of the cogeneration system. Indices: i is the number of elements in the cogeneration system; 0, 1, 2 are the initial stationary mode, external and internal character of influences.

The mathematical substantiation of maintenance of the operation of the biogas cogeneration system Smart Grid (2), (Figure 2.1), based on the methodology of the mathematical description of dynamics of power systems, the method of the graph of cause-effect relations [Chapter 1] is proposed. The basis of the proposed rationale is the mathematical description of the architecture of the biogas cogeneration system Smart (1), (Figure 2.1). Prediction of a change in power factor and a change in the temperature of local water of the engine cooling circuit makes it possible to make forestalling decisions on a change of the heat pump power and the number of heat exchanger plates of the engine cooling circuit. The voltage at the inlet of the inverter, at the outlet of the inverter and voltage frequency are measured. In the engine cooling circuit, the temperature of the cooling water at the inlet of the heat exchanger, at the outlet of the heat exchanger, and the temperature of return water are measured.

Mathematical substantiation of Smart Grid maintenance of the operation of the biogas cogeneration system (2) is proposed (Figure 2.1).

Where $BGCSSG(\tau)$ is the Smart Grid maintenance of the operation of the cogeneration biogas system; τ is the time, s; $ID(\tau)$ is the integrated dynamic subsystem (the electric network, the cogeneration unit, the engine cooling circuit as a part of the cogeneration plant, the biogas plant, the heat pump, the frequency converter); $P(\tau)$ is the properties of the elements of the integrated dynamic subsystem, units of the cogeneration system; $CMM(\tau, z)$ is the complex mathematical modeling of dynamics of fermentation temperature, the consumption of refrigerant vapor, the number of revolutions of the compressor electric motor, the temperature of the heating water at the outlet of the condenser of the heat pump, a change in power factor, the temperature of local water; $sd(\tau)$ is the input data (productivity of the biogas plant and the type of the cogeneration system and its power, the heat pump type and its power, the integrated system of maintenance of fermentation temperature, the frequency converter type; $lp(\tau)$ is the boundary change in parameters (the temperature of heating water at the outlet from the heat exchanger fitted in the methane tank, the temperature of the refrigerant at the inlet to the condenser and at the outlet of the condenser, the temperature of the wort at the outlet of the evaporator, voltage at the inlet of the inverter and at the outlet of the inverter, voltage frequency, temperature of cooling water at the inlet of the heat exchanger and at the outlet of the heat exchanger, the temperature of return water; $lf(\tau)$ is the levels of operation of biogas plant, heat pump system, of the cogeneration system; $fd(\tau)$ is the obtained parameters (parameters of heat exchange in the heat exchanger fitted in the methane tank, heat pump system, engine cooling circuit, parameters of heat exchange in the heat exchanger of the engine cooling circuit, time constants and coefficients of the mathematical model of dynamics of a change in fermentation temperature, the temperature of the heating water at the outlet of the condenser of the heat pump, power factor, the temperature of local water; $tf(\tau,z)$ is the transfer function of predicted parameters – fermentation temperature, in the consumption of refrigerant vapor, the number of revolutions of the compressor motor, the temperature of the heating water at the outlet of the condenser of the heat pump, power factor, the temperature of local water; $AI(\tau,z)$ is the standard information regarding the evaluation of the maximum admissible change in fermentation temperature, in the consumption of refrigerant vapor, the number of revolutions of the compressor motor the temperature of local water; $C(\tau)$ is the control of workability of the cogeneration system; $LC(\tau)$ is the logical relations of temperature of the heating water at the outlet of the condenser of the heat pump the temperature of the heating water at the outlet of the condenser of the heat pump, power factor, the control of the cogeneration system workability; $x(\tau)$ is the influences (a change in parameters: the temperature of heating water at the outlet from the heat exchanger fitted in the methane tank, the temperature of refrigerant at the outlet from the condenser, evaporation pressure, condensation pressure, voltage at the inlet to the inverter and at the outlet from the inverter, voltage frequency, temperature of cooling water at the inlet to the heat exchanger, at the outlet from the heat exchanger, the temperature of return water; $f(\tau)$ is the measured parameters (the temperature of heating water at the outlet from the heat exchanger fitted in the methane tank, the temperature of refrigerant at the outlet from the condenser, evaporation pressure, voltage at the inlet of the inverter and at the outlet of the inverter, voltage frequency, the temperature of cooling water at the inlet of the heat exchanger, at the outlet of the heat exchanger, the temperature of return water); $K(\tau)$ is the coefficients of the mathematic description of the dynamics of a change in the temperature of fermentation, the consumption of refrigerant vapor, the number of revolutions of the compressor electric motor, the temperature of heating water at then outlet from the condenser, power factor, the temperature of local water; $y(\tau, z)$ is the output parameters (the temperature of fermentation, the consumption of refrigerant vapor, the number of revolutions of the compressor electric motor, the temperature of heating water at then outlet from the condenser, power factor, reactive power factor, active, reactive powers of the cogeneration system, temperature of local water); z is the coordinate of the length: the heat exchanger fitted in the methane tank, heat pump condenser, of the heat exchanger of the engine cooling circuit, *m*; $d(\tau)$ is the dynamic parameters of a change in the temperature of fermentation, the consumption of refrigerant vapor, the number of revolutions of the compressor electric motor, the temperature of heating water at the outlet from the heat pump condenser, the power factor of the cogeneration system, the temperature of local water; $FI(\tau)$ is the functional final information regarding decision making; $LMD(\tau)$ is the logical relations of decision making; $MD(\tau)$ is decision making; $NC(\tau)$ is the new conditions of the cogeneration system operation; $S(\tau)$ is the identification of the state of the cogeneration system; $LS(\tau)$ is the logical relations of identification of the state of the cogeneration system; $R(\tau)$ is the logical relations between the dynamic subsystem and units of charge, discharge, functional estimation of efficiency that belong to the cogeneration system. Indices: i is the number of elements of $SGCBS$ (τ); 0, 1, 2 are the initial, external, and internal character of influences.

Mathematical substantiation of the architecture of the biogas cogeneration system Smart (1) and mathematical substantiation of maintenance of the operation of the biogas cogeneration system Smart Grid (2) (Figure 2.1) make it possible to maintain the operation of the cogeneration system using the following actions:

- Workability control ($C(\tau)$) of the dynamic subsystem based on complex mathematical ($CMM(\tau, z)$) and logical ($LC(\tau)$) modeling regarding obtaining standard ($AI(\tau,z)$) estimate of a change in the temperature of fermentation, in the consumption of refrigerant vapor, the number of revolutions of the compressor electric motor, the temperature of heating water at outlet from the condenser, power factor of the cogeneration system, of a change in the temperature of local water;

- Workability control ($C(\tau)$) of the dynamic system based on complex mathematical ($CMM(\tau, z)$) and logical ($LC(\tau)$) modeling regarding the obtaining functional ($FI\ (\tau)$) estimate of a change in temperature of fermentation, in the consumption of refrigerant vapor, the number of revolutions of the compressor electric motor, the temperature of heating water at the outlet from the condenser, power factor of the cogeneration system, the temperature of local water;
- Decision making ($MD(\tau)$) with the use of the functional resulting information ($FI\ (\tau)$), obtained based on logical modeling ($LMD(\tau)$);
- Decision making to maintain the temperature of fermentation, in the consumption of refrigerant vapor, the number of revolutions of the compressor electric motor, the temperature of heating water at then outlet from the condenser, ratio of production and consumption of electric power and heat with the use of the functional assessment of a change in power factor, the temperature of local water, with the use of the functional resulting information ($FI\ (\tau)$);
- Identification ($S(\tau)$) of the new conditions of functioning of the cogeneration system ($NC(\tau)$) based on logical modeling ($LS(\tau)$) as a part of the dynamic subsystem and confirmation of new operating conditions based on logical modeling ($R(\tau)$) from the units of the cogeneration system.

Smart System to Support the Operation of the Biogas Plant at the Decision-Making Level

The most important in anaerobic fermentation is the productive activity of the biogas production process, the maintenance of which within the specified limits affects the qualitative and quantitative return of the methane tank. Moreover, there is a need to ensure the marketability of the biogas plant, taking into account energy losses to the environment, to heat the substrate to the temperature of fermentation and mixing. The operation of biogas plants is accompanied by inconsistent output of biogas due to the difficulty of determining the exact timing of changes in the flow of fresh and fermented raw materials, which requires additional equipment. In the conditions of unreliable use of the estimation of the change of fermentation temperature during the measurement, which is due to the significant thermal accumulating capacity of the wort, the cost of supporting the fermentation process is up to 20-30% of the energy produced. Due to the fact that the temperature of fresh wort loaded depends on the ambient temperature, the change in wort consumption to support the fermentation process disrupts the mandatory balance of fresh and fermented raw material flows to achieve a constant biogas yield. The use of changes in the consumption of the heating medium to support the fermentation process may adversely affect the activity of the fermentation process.

It is proposed to support the operation of the biogas plant based on integrated analytical assessment of changes in fermentation temperature at the decision-making level to support the main components of the biogas production process: biogas yield, shipment of fermented wort and loading of fresh material. For timely decision-making, it is proposed to predict the change in fermentation temperature by measuring the temperature of the heating coolant at the outlet of the heat exchanger built into the methane tank, which changes earlier than the change in fermentation temperature.

It is proposed to set the temperature of the heating coolant at the inlet to the heat exchanger using a heat pump power supply, for which the low-potential energy source is fermented wort. The prediction of a change in the fermentation temperature requires making advanced decisions about a change in the temperature of the warming heat carrier at the inlet to the heat exchanger, fitted in the methane tank based on the change of the number of rotations of the electric motor of the heat pump spiral compressor.

According to formulas (1), (2), the prediction of a change in the temperature fermentation is proposed. The temperature of heating water at the outlet from the heat exchanger fitted in the methane tank is measured. The transfer function along the channel "temperature fermentation – temperature of heating water at the outlet of the heat exchanger fitted in the methane tank" is obtained. A change in the fermentation temperature is estimated both over time and along the spatial coordinate of the heat exchanger coincides with the direction of the flow of motion of the medium. The transfer function along the channel "temperature fermentation – temperature of heating water at the outlet of the heat exchanger fitted in the methane tank," which was obtained as a result of solving a system of nonlinear differential equations, is presented as follows:

$$W_{t-\vartheta_1} = \frac{K_{hw}\varepsilon\left(1-L_{hw}^{*}\right)}{\left(T_{wr}S+1\right)\beta-1}\left(1-e^{-\gamma\xi}\right), \tag{3}$$

where

$$K_{hw} = \frac{m\left(\theta_0-\sigma_0\right)}{G_{hw0}};\quad \varepsilon = \frac{\alpha_{hw0}h_{hw0}}{\alpha_{wr0}h_{wr0}};\quad L_{hw} = \frac{G_{hw}C_{hw}}{\alpha_{hw0}h_{hw0}};$$

$$T_{wr} = \frac{g_{wr}C_{wr}}{\alpha_{wr0}h_{rw0}};\quad \beta = T_m S+\varepsilon^{*}+1;\quad T_{M} = \frac{g_m C_m}{\alpha_{wr0}h_{wr0}};\quad \varepsilon^{*} = \varepsilon\left(1-L_{hw}^{*}\right);$$

$$\gamma = \frac{\left(T_{wr}S+1\right)\beta-1}{\beta};\quad \xi = \frac{z}{L_{wr}};\quad L_{wr} = \frac{G_{wr}C_{wr}}{\alpha_{wr0}h_{wr0}},$$

where C is the specific thermal capacity, kJ/(kg·K); α is the heat transfer factor, kW/(m^2·K); G is the consumption of substance, kg/s; g is the specific weight of a substance, kg/m; h is the specific surface, m^2/m; t, σ, θ are the temperature of the fermentation, heating water from the outlet from the heat exchanger fitted in the methane tank and of the separating wall, respectively, K; z is spatial coordinate of the heat exchanger coincides with the direction of the flow of motion of the medium, m; $T_{wt,}$, T_m are the time constants that characterize the thermal accumulating capacity of wort, metal, s; m is the indicator of the dependence of heat transfer factor on consumption; τ is the time, s; S is the Laplace parameter; $S = \omega j$; ω is the frequency, 1/s.; Indices: 0 – initial stationary mode; 1 – inlet to the heat exchanger; wt – internal flow – wort; hw – external flow – heating water ; m – metal wall.

Transfer function along the channel: "temperature of fermentation – temperature of heating water at the outlet of the heat exchanger fitted in the methane tank" was obtained based on the solution of a system of nonlinear differential equations using the Laplace transform tool. The system of differential equations includes the equation of state as the estimation of the physical model of the biogas plant. The system of differential equations also includes the equation of energy of transmitting and receiving media – heating water and wort, respectively, and the equation of thermal balance for the wall of the heat exchanger. The equation of the energy of the receiving medium is developed with the representation of a change in temperature of the fermentation both in time and along the spatial coordinate, which coincides with the direction of the motion flow of the medium. The equation of the energy of the transmitting medium includes the K_{hw} coefficient, which assesses a change in the temperature of heating water at the outlet of the heat exchanger fitted in the methane tank that is measured.

A real part of the transfer function was separated:

$$O(\omega)=\frac{(L_1A_1)+(M_1B_1)(1-L_{hw}^{*})}{(A_1^2+B_1^2)}. \tag{4}$$

The K_{hw} factor includes the temperature of the separating wall θ:

$$\theta=(\alpha_{B}(\sigma_1+\sigma_2)/2)+(A(t_1+t_2)/2)/(\alpha_{wr}+A), \tag{5}$$

where σ_1, σ_2 are the temperatures of heating water at the inlet and at the outlet of the heat exchanger fitted in the methane tank, K, respectively; t_1, t_2 are the wort temperatures at the inlet and at the outlet of the heat exchanger fitted in the methane tank, K, respectively; α is the heat transfer factor, kW/(m^2·K). Index wr – internal flow: wort.

$$A=1/(\delta_m/\lambda_m+1/\alpha_{hw}), \tag{6}$$

where δ is the thickness of a wall of the heat exchanger, m; λ is the thermal conductivity of the metal wall of the heat exchanger, kW/(m·K). Indices: hw – external flow: heating water; *m* – metal wall of a heat exchanger.

To use the real part $O(\omega)$, the following factors were obtained:

$$A_1=\varepsilon^*-T_{wr}T_m\omega^2; \tag{7}$$

$$A_2=\varepsilon^*+1; \tag{8}$$

$$B_1 = T_{wr}\varepsilon^*\omega + T_{wr}\omega + T_m\omega; \tag{9}$$

$$B_2 = T_m\omega; \tag{10}$$

$$C_1 = \frac{A_1A_2 + B_1B_2}{A_2^{\;2} + B_2^{\;2}}; \tag{11}$$

$$D_1 = \frac{A_2B_1 - A_1B_2}{A_2^{\;2} + B_2^{\;2}}; \tag{12}$$

$$L_1 = 1 - e^{-\zeta C_1}\cos\left(-\xi D_1\right); \tag{13}$$

$$M_1 = -e^{-\zeta C_1}\sin\left(-\xi D_1\right). \tag{14}$$

The transfer function (3), which was obtained based on the use of the operator method of solving the system of nonlinear differential equations, retains the Laplace transform parameter – S ($S = \omega j$), where ω is the frequency, 1/s. To switch from the frequency area to the time area, a real part (4), obtained as a result of the mathematical treatment of transfer function, was separated.

It is this part that is included in the integrals (15), which makes it possible to obtain dynamic characteristics of a change the fermentation temperature using the inverse Fourier transform:

$$t(\tau, z) = \frac{1}{2\pi}\int_0^\infty K_{hw}O(\omega)\sin\left(\tau\omega/\omega\right)d\omega, \tag{15}$$

where t is the fermentation temperature, K.

So, for obtaining the reference estimation of change in the fermentation temperature a block diagram (Figure 2.2) is proposed using, for example, the initial data of constructive-regime implementation of a methane tank producing 352.5 m^3/day of biogas.

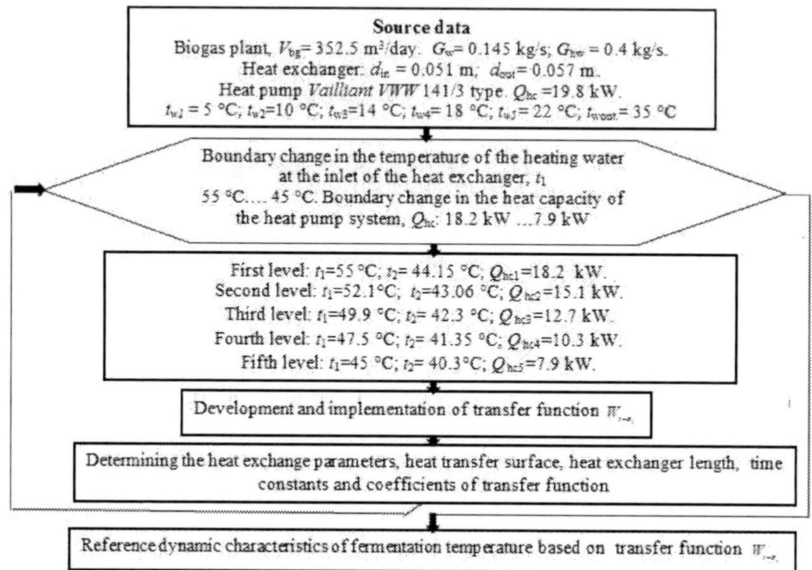

Figure 2.2. Block diagram of complex mathematical modeling of the biogas plant and the heat exchanger fitted in the methane tank: V_{bg} – productivity of biogas plant, m^3/day; G_w, G_{hw} – consumption of wort, heating water, kg/s, respectively; d_{in}, d_{out} – diameter of the heat exchanger, inner, outer, m, respectively; Q_{hc} – heat capacity of the heat pump system, kW; t_{w1}, t_{w2}, t_{w3}, t_{w4}, t_{w5} – the temperature of the wort at the inlet to the biogas plant depending on the ambient temperature, °C, respectively; $t_{w\,out}$ – temperature of the wort at the outlet of the biogas plant, °C; t_1, t_2 – temperature of the heating water at the inlet and the outlet of the heat exchanger fitted in the methane tank, °C, respectively.

The biogas plant with a capacity of 325 m^3/day of biogas production includes the following components: the heat exchanger fitted in the methane tank with outer diameter, inner diameter 0,057 m, 0,051 m, respectively. The heat pump system type, Vailliant VWW 141/3 (Germany) with a heating capacity of 19.8 kW. Boundary change of heating water temperature at the inlet of heat exchanger fitted in the methane tank: 55…45°C, boundary change of the heat load of the heat pump system: 18.2…7.9 kW. The following levels of operation of the biogas plant have been established for the change in the temperatures of heating water at the inlet of heat exchanger fitted in the methane tank and at the outlet of heat exchanger fitted in the methane tank, respectively: – first level: 55°C – 44.15°C; second level: 52.1°C – 43.06°C; third level: 49.9°C – 42.3°C; fourth level: 47.5°C –41.35°C; fifth level: 45°C –40.3°C. corresponding to changes in the temperature of wort at the inlet biogas plant: 5°C, 10°C, 14°C, 18°C , 22°C and fermentation temperature 34... 36°C.

According to formulas (1) - (3), block diagram of complex mathematical modeling (Figure 2.1) the results of complex mathematical modeling of the biogas plant and the heat exchanger fitted in the methane tank are presented (Tables 2.1, 2.2).

Table 2.1. Heat transfer parameters as part of complex mathematical modeling of methane tank

Levels of operation	Parameter		
	α_{hw}, kW/(m²·K)	α_{wr}, kW/(m²·K)	k, kW/(m²·K)
First level	1.158	0.631	0.398
Second level	1.135	0.615	0.389
Third level	1.125	0.595	0.380
Fourth level	1.114	0.575	0.371
Fifth level	1.102	0.549	0.358

Note: α_{hw} – coefficient of convective heat transfer from the heat water to the heat exchanger wall, kW/(m²·K); α_{wr} – coefficient of convective heat transfer from the heat exchanger wall to wort, kW/(m²·K); k – heat transfer coefficient, kW/(m²·K).

Table 2.2. Time constants and coefficients of the mathematical model of the dynamics of the temperature of fermentation

Levels of operation	T_{wr}, c	T_m, c	ε	ε^*	ζ	L_{wr}, m	L_{hw}, m	L_{hw}^*
First level	99. 57	14.75	1.63	1.47	1.76	5.36	9.11	0.099
Second level	102.23	15.15	1.65	1.49	1.72	5.51	9.22	0.098
Third level	105.54	15.64	1.69	1.53	1.66	5.69	9.30	0.097
Fourth level	109.25	16.19	1.73	1.57	1.60	5.89	9.39	0.096
Fifth level	114.49	16.96	1.8	1.63	1.53	6.17	9.49	0.095

Presented in Table 2.2, time constants and coefficients that are part of the mathematical models of dynamics (3) are obtained on the basis of parameters as part of complex mathematical modeling (Table 2.1).

The proposed block diagram allows, obtaining the time constants and coefficients of the mathematical model of the dynamics of fermentation temperature, to determine the boundary change in fermentation temperature for the established levels of operation.

Based on the proposed mathematical substantiation Smart Grid maintenance of functioning of the cogeneration system (1) – (3), the block diagram for the control of efficiency of the cogeneration on biogas fuel (Figure 2.3) is developed.

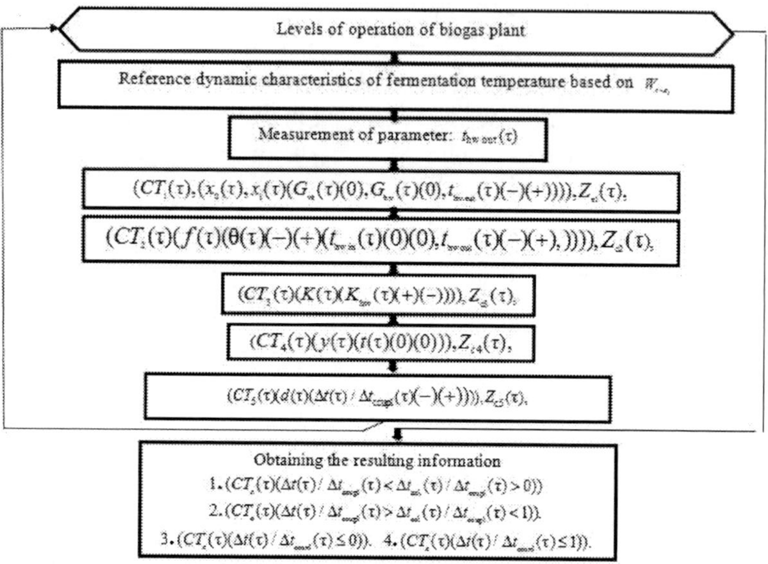

Figure 2.3. Block diagram of control of efficiency of the biogas plant: $t_{hw\ in}$, $t_{hw\ out}$ are the temperatures of heating water at the inlet to the heat exchanger and the outlet of the heat exchanger, K, respectively; *CT* is the event control; *x* is the influences; G_{wt}, G_{hw} are the consumption of wort, heating water, kg/s, respectively; *Z* is the logical relations; *f* is the diagnosed parameters; θ is the temperatures of the separating wall of the heat exchanger, K; *K* is the mathematic description; *t* is fermentation temperature, K; *y* is the output parameter;. τ – час, s. Indices: *c* is efficiency control; hw is heating water; ccupl, ccl are constant, calculated value of the parameter of the upper level of functioning, level of functioning, respectively; 0, 1, 2 are the initial stationary mode, external, internal influences; 3 is the coefficients of dynamics equations; 4 is the essential predicted parameters; 5 is dynamic parameters.

Control of efficiency of the biogas plant (Figure 2.3) makes it possible to obtain the resulting information for making a decision on maintaining the functioning of a biogas plant. Based on the proposed mathematical substantiation (1) - (3), the block diagram (Figure 2.3) of the maintenance of the functioning of the biogas plant is developed based on the maintenance of change in functional efficiency.

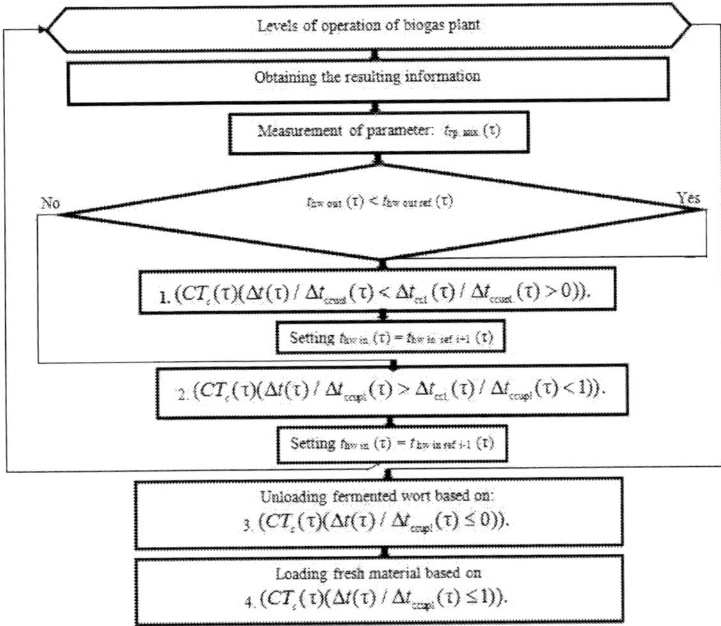

Figure 2.4. Block diagram of support for the functioning of the biogas plant at the decision-making level: *CT* is the event control; *t* is fermentation temperature, K; $t_{hw\ in}$, $t_{hw\ out}$ are the temperatures of heating water at the inlet to the heat exchanger and the outlet of the heat exchanger, K, respectively; $t_{hw\ in\ ref}$, $t_{hw\ out\ ref}$ are reference values of temperatures of heating water at the inlet to the heat exchanger and the outlet of the heat exchanger, K, respectively; τ – time, s. Indices: *c* is efficiency control; hw is heating water; ccupl, ccl are constant, calculated value of the parameter of the upper level of functioning, level of functioning, respectively.

The temperature of the heating water at the outlet of the heat exchanger built into the methane tank, that is measured, is compared with the reference value of the operating level. The use of a logical structure within the cycle provides an opportunity to maintain the discharge or charge of the biogas plant, accompanied by the output of biogas, based on the obtained information assessment (1), (2), respectively (Figure 2.3). Obtaining the final information (3) requires a decision on the unloading of fermented wort, which is associated with reaching the temperature of the heating water at the outlet of the extremely low heat exchanger – 37.04°C, Loading fresh material based on the final assessment (4) requires setting the temperature of the heating water at the inlet to the heat exchanger built into the methane tank at the upper limit of – 55°C. The comprehensive integrated system of maintenance of operation of the fermentation temperature is developed. (Table 2.3). There is a continuous measurement of the temperature of heating water at the outlet of the heat exchanger fitted in the methane tank is measured. The proposed system allows to charge the cogeneration system during the discharge period and charge of the biogas plant for biogas output and to set the exact time of unloadin of fermented material and loading of fresh wort to ensure the support of the cogeneration unit.

Table 2.3. Integrated Smart Grid system of maintenance of fermentation temperature

Time, τ, 100 s	Change in fermentation temperature	$\Delta t(\tau)/\Delta t(\tau)_1$	$t(\tau)$, °C
13	Loading fresh material Setting temperature t_{in} =55°C	1	36
26	Discharge – charge; t_{out} =43.6°C	0.8874	35.77
39	Decision making Setting temperature t_{in} =52.1°C.	0.8866	35.77
52	Discharge – charge; t_{out} =42.6°C	0.8130	35.62
65	Decision making Setting temperature t_{in} =49.9°C.	0.8119	35.62
78	Discharge – charge; t_{out} =41.5°C	0.6871	35.37
91	Decision making Setting temperature t_{in} =47.5°C.	0.6823	35.36
104	Discharge – charge; t_{out} =40°C	0.4872	34.97
117	Decision making. Setting temperature t_{in} =45°C	0.4870	34.97
130	t_{out} =37.04°C. Unloading fermented wort	0	34

Note: t_{in} is set temperatures of heat water at the inlet of the heat exchanger fitted in the methane tank, °C; t_{out} is measured temperatures of heat water at the outlet of the heat exchanger fitted in the methane tank, °C; t is the fermentation temperature, °C; Index: 1 – established calculation value of the parameter of the upper level of operation.

$$t_{i+1}(\tau) = t_i - \left(\begin{array}{l} \Delta t_i(\tau) / \Delta t_1(\tau) - \\ -\Delta t_{i+1}(\tau) / \Delta t_1(\tau) \end{array} \right) (t_1 - t_2), \tag{16}$$

where t is the temperature of fermentation, °C; t_1, t_2 are the initial, final values of temperature of fermentation; τ is the time, s. Index: 1 is the constant calculation value of the parameter of the upper level of operation; i is the number of levels of functioning of the biogas plant.

Thus, for example, in the time interval $65\cdot 10^2$ s (1.8 hours) from the loading of fresh material into the biogas plant (Table 2.3), the decision was made to set the temperature of the warming heat carrier at the inlet of the heat exchanger, fitted in a methane tank, at the level of 49.9°C. The temperature of the fermentation of raw material is maintained at the level of 35.62°C (Table 2.3).

35.62°C=35.62°C- (0.8130–0.8119)(36–34°C).

Thus, for example, in the time interval $91\cdot 10^2$ s (2.53 hours) from the loading of fresh material into the biogas plant (Table 2.3), the decision was made to set the temperature of the warming heat carrier at the inlet of the heat exchanger, fitted in a methane tank, at the level of 47.5°C. The temperature of the fermentation of raw material is maintained at the level of 35.36°C (Table 2.3).

35.36°C=35.37°C- (0.6871–0.6823)(36–34°C).

If in the time interval $130\cdot 10^2$ s (3.61 hours) temperature of the heating water at the outlet of the heat exchanger, fitted in a methane tank, decreased at the level of 37.4°C corresponding to the fermentation temperature 34°C, it is necessary unload fermented wort. To load fresh material, it is necessary to set the temperature of the heating water at the inlet to the heat exchanger at 55°C.

The proposed integrated system to support the operation of the biogas plant, which on the basis of integrated analytical assessment of fermentation temperature allows to set the temperature of the heating water at the inlet to the heat exchanger built into the methane tank when measuring the temperature of the heating water at the outlet of the heat exchanger and perform timely unloading of fermented wort and loading of raw material.

Integrated Smart Grid Fermentation Temperature Support System Using Heat Pump Power Supply

Support for the operation of the heat pump power supply of the biogas plant is based on forecasting changes in the temperature of the heating water at the inlet to the heat exchanger built into the methane tank. Advance decisions on changing the number of revolutions of the compressor electric motor to change the refrigerant consumption occurs when measuring the temperature of the refrigerant at the outlet of the condenser, the evaporation pressure at the outlet of the evaporator, the condensation pressure at the outlet of the condencer to establish their ratio, the temperature of the heating water at the outlet of the heat exchanger built into the methane tank and voltage frequency.

According to formulas (1), (2), the prediction of a change in the temperature of heating water at the outlet of the condencer for the heat exchanger fitted in the methane tank was proposed. The temperature of heating water at the inlet at the condencer from heat exchanger fitted in the methane tank is measured.

The transfer function along the channel "temperature of heating water at the outlet of the condencer– temperature of heating water at the inlet of the condencer" was obtained. A change in the temperature of heating water at the condencer is estimated both over time and along the spatial coordinate of the condencer coincides with the direction of the flow of motion of the medium. The transfer function along the channel " temperature of heating water at the outlet of the condencer– temperature of heating water at the inlet of the condencer," which was obtained as a result of solving a system of nonlinear differential equations, is presented as follows:

$$W_{t-Gr1} = \frac{K_r \varepsilon \left(1 - L_r^*\right)}{(T_{hw} S + 1)\beta - 1}\left(1 - e^{-\gamma\xi}\right), \tag{17}$$

where

$$K_r = \frac{m\left(\theta_0 - \sigma_0\right)}{G_{r0}}; \quad \varepsilon = \frac{\alpha_{r0} h_{r0}}{\alpha_{hw0} h_{hw0}};$$

$$L_r^* = \frac{1}{L_r + 1}; \quad L_r = \frac{G_r C_r}{\alpha_{r0} h_{r0}}; \quad \gamma = \frac{(T_{hw} S + 1)\beta - 1}{\beta};$$

$$T_{hw} = \frac{g_{hw} C_{hw}}{\alpha_{hw0} h_{hw0}}; \; \xi = \frac{z}{L_{hw}}; \; L_{hw} = \frac{G_{hw} C_{hw}}{\alpha_{hw0} h_{hw0}};$$

$$\beta = T_m S + \varepsilon^* + 1; \ \varepsilon^* = \varepsilon(1\text{-}L_r^*); \ T_m = \frac{g_m C_m}{\alpha_{hw0} h_{hw0}},$$

where C is the specific thermal capacity, kJ/(kg·K); α is the heat transfer factor, kW/(m^2·K); G is the consumption of substance, kg/s; g is the specific weight of a substance, kg/m; h is the specific surface, m^2/m; t, σ, θ – temperature of heating water, temperature of refrigerant and of the separating wall, respectively, K; z is spatial coordinate of the condenser coincides with the direction of the flow of motion of the medium, m; $T_{hw,}$, T_m are the time constants that characterize the thermal accumulating capacity of heating water, metal, s; m is the indicator of the dependence of heat transfer factor on consumption; τ is the time, s; S is the Laplace parameter; $S = \omega j$; ω is the frequency, 1/s.; Indices: 0 – initial stationary mode; 1 – inlet to the condenser; hw – internal flow – heating water; r – external flow – refrigerant; m – metal wall.

Transfer function along the channel: "temperature of heating water at the outlet of the condenser– temperature of heating water at the inlet of the condenser was obtained based on the solution of a system of nonlinear differential equations using the Laplace transform tool. The system of differential equations includes the equation of state as the estimation of the physical model of the biogas plant – heat pump system. The system of differential equations also includes the equation of energy of transmitting and receiving media – refrigerant and local water, respectively, and the equation of thermal balance for the wall of the condenser. The equation of the energy of the receiving medium is developed with the representation of a change in temperature of the local water both in time and along the spatial coordinate, which coincides with the direction of the motion flow of the medium. The equation of the energy of the transmitting medium includes the K_r coefficient, which assesses a change in the temperature of heating water at the inlet of the condenser, the temperature of refrigerant at the outlet from the condenser that are measured at the established change of temperature of the heated water.

According to formulas (1), (2), the prediction of a change in refrigerant consumption was proposed. The evaporation pressure of the refrigerant at the outlet of the evaporator and condensing pressure of the refrigerant at the outlet of the condenser are measured. The transfer function along the channel "refrigerant consumption – refrigerant pressure" as a result of solving a system of nonlinear differential equations, using the Laplace transform tool. The system of differential equations includes the equation of state as the estimation of the physical model of the biogas plant – heat pump system. The system of differential equations also includes the equation of energy of transmitting and receiving media – refrigerant and heating water, respectively, the equation of continuity and the equation of thermal balance for the wall of the condenser.

The transfer function along the channel "refrigerant consumption – refrigerant pressure" is presented as follows:

$$W_{Gr-p_1} = \frac{\chi_p S}{\gamma}\left(1 - e^{-\gamma_1 \xi}\right), \tag{18}$$

where

$$\chi_p = -f_s \frac{\partial \rho}{\partial p}; \quad \gamma = \frac{(T_{hw} S + 1)\beta - 1}{L_{hw}\beta}; \quad \gamma_1 = \frac{(T_{hw} S + 1)\beta - 1}{\beta};$$

$$T_{hw} = \frac{g_{hw} C_{hw}}{\alpha_{hw0} h_{hw0}}; \quad L_{hw} = \frac{G_{hw} C_{hw}}{\alpha_{hw0} h_{hw0}};$$

$$\beta = T_m S + \varepsilon^* + 1; \quad T_m = \frac{g_m C_m}{\alpha_{hw0} h_{hw0}};$$

$$\varepsilon^* = \varepsilon(1 - L_r^*); \quad \varepsilon = \frac{\alpha_{r0} h_{r0}}{\alpha_{hw0} h_{hw0}};$$

$$L_r^* = \frac{1}{L_r + 1}; \quad L_r = \frac{G_r C_r}{\alpha_{r0} h_{r0}}; \quad \xi = \frac{z}{L_{hw}},$$

where α is the heat transfer factor, kW/(m^2·K); p - refrigerant pressure, MPa; f_s – cross section for the passage of refrigerant, m^2; C is the specific thermal capacity, kJ/(kg·K); G is the consumption of substance, kg/s;; ρ – refrigerant density, kg/m^3; g is the specific weight of a substance, kg/m; h is the specific surface, m^2/m; z is spatial coordinate of the condenser coincides with the direction of the flow of motion of the medium, m; T_{hw},, T_m are the time constants that characterize the thermal accumulating capacity of heating water, metal, s; m is the indicator of the dependence of heat transfer factor on consumption; τ is the time, s; S is the Laplace parameter; $S = \omega j$; ω is the frequency, 1/s.; Indices: 0 – initial stationary mode; 1 – inlet to the con-denser; hw – internal flow – heating water; r – external flow – refrigerant; m – metal wall.

According to formulas (1), (2), the prediction of a change in the number of revolutions of the electric motor of the heat pump compressor was proposed. The evaporation pressure of the refrigerant at the outlet of the evaporator, condensing pressure of the refrigerant at the outlet of the condenser and voltage frequency are measured. The transfer function along the channel: "number of revolutions of the electric motor of the compressor of the heat pump - voltage frequency" is presented as follows:

$$W_{n-f_1} = \frac{K_f \chi_p S}{\gamma}\left(1 - e^{-\gamma_1 \xi}\right), \qquad (19)$$

$$K_f = \frac{60 f (1 - s)}{p_n}; \quad \chi_p = -f_s \frac{\partial \rho}{\partial p}; \quad \gamma = \frac{(T_{\text{hw}} S + 1)\beta\text{-}1}{L_{\text{hw}} \beta};$$

$$\gamma_1 = \frac{(T_{\text{hw}} S + 1)\beta - 1}{\beta}; \; T_{\text{hw}} = \frac{g_{\text{hw}} C_{\text{hw}}}{\alpha_{\text{hw}0} h_{\text{hw}0}}; \; L_{\text{hw}} = \frac{G_{\text{hw}} C_{\text{hw}}}{\alpha_{\text{hw}0} h_{\text{hw}0}};$$

$$\beta = T_{\text{m}} S + \varepsilon^* + 1; \; T_{\text{m}} = \frac{g_{\text{m}} C_{\text{m}}}{\alpha_{\text{hw}0} h_{\text{hw}0}}; \; \varepsilon^* = \varepsilon(1 - L_{\text{r}}^*);$$

$$\varepsilon = \frac{\alpha_{\text{r}0} h_{r0}}{\alpha_{\text{hw}0} h_{\text{hw}0}}; \; L_r^* = \frac{1}{L_{\text{r}} + 1}; \; L_{\text{r}} = \frac{G_{\text{r}} C_{\text{r}}}{\alpha_{\text{r}0} h_{\text{r}0}}; \; \xi = \frac{z}{L_{\text{hw}}},$$

where α is the heat transfer factor, kW/(m^2·K); p is refrigerant pressure, MPa; f_s is cross section for the passage of refrigerant, m^2; f is the voltage frequency, Hz; p_n is the number of pole pairs; s is amount of slip of the electric motor; C is the specific thermal capacity, kJ/(kg·K); G is the consumption of substance, kg/s;; ρ is refrigerant density, kg/m^3; g is the specific weight of a substance, kg/m; h is the specific surface, m^2/m; z is spatial coordinate of the condenser coincides with the direction of the flow of motion of the medium, m; T_{hw},, T_{m} are the time constants that characterize the thermal accumulating capacity of heating water, metal, s; m is the indicator of the dependence of heat transfer factor on consumption; τ is the time, s; S is the Laplace parameter; $S = \omega j$; ω is the frequency, 1/s.; Indices: 0 – initial stationary mode; 1 – inlet to the condenser; hw – internal flow – heating water; r – external flow – refrigerant; m – metal wall.

A real part of the transfer function (17) was separated:

$$O_1(\omega) = \frac{(L_1 A_1) + (M_1 B_1) K_r \varepsilon (1 - L_3^*)}{(A_1^2 + B_1^2)}. \tag{20}$$

The K_r factor includes the temperature of the separating wall θ:

$$\theta = (\alpha_{hw}(\sigma_1 + \sigma_2)/2) + A(t_1 + t_2)/2)/(\alpha_{hw} + A), \tag{21}$$

where σ_1, σ_2 are the temperatures of refrigerant at the inlet and the outlet of the condenser, respectively, K; t_1, t_2 are the temperatures of heating water at the inlet and the outlet of the condenser, respectively, K; α is the heat transfer factor, kW/(m^2·K); Indices: hw is internal flow: heating water.

$$A = 1/(\delta_m/\lambda_m + 1/\alpha_r), \tag{22}$$

where δ is the thickness of a wall of the condenser, m; λ is the thermal conductivity of the metal wall of the condenser, kW/(m·K). α is the heat transfer factor, kW/(m^2·K); Indices: r – external flow: refrigerant; m – metal wall of a condenser.

To use the real part $O_1(\omega)$, the following factors were obtained:

$$A_1 = \varepsilon^* - T_{hw} T_m \omega^2; \tag{23}$$

$$A_2 = \varepsilon^* + 1; \tag{24}$$

$$B_1 = T_{hw}\varepsilon\omega + T_{hw}\omega + T_m\omega; \tag{25}$$

$$B_2 = T_m\omega; \tag{26}$$

$$C_1 = \frac{A_1 A_2 + B_1 B_2}{A_2^{\ 2} + B_2^{\ 2}}; \tag{27}$$

$$D_1 = \frac{A_2 B_1 - A_1 B_2}{A_2^{\ 2} + B_2^{\ 2}}; \tag{28}$$

$$L_1 = 1 - e^{-\xi C_1} \cos(-\xi D_1); \tag{29}$$

$$M_1 = -e^{-\xi C_1} \sin(-\xi D_1). \tag{30}$$

A real part of the transfer function (18) was separated:

$$O_2(\omega) = \chi_p L_{\mathrm{hw}} (C_1 L_1) \text{-} (D_1 M_1). \tag{31}$$

To use the real part $O_2(\omega)$, the following factors were obtained:

$$A_1 = -T_{\mathrm{m}} \omega^2; \tag{32}$$

$$A_2 = \varepsilon \ - T_{\mathrm{hw}} T_{\mathrm{m}} \omega^2; \tag{33}$$

$$B_1 = (\varepsilon + 1)\omega; \tag{34}$$

$$B_2 = T_{hw} \varepsilon\omega + T_{\mathrm{hw}} \omega + T_{\mathrm{m}} \omega + \varepsilon, \tag{35}$$

$$C_1 = \frac{A_1 A_2 + B_1 B_2}{A_2^{\ 2} + B_2^{\ 2}}; \tag{36}$$

$$D_1 = \frac{A_2 B_1 - A_1 B_2}{A_2^{\ 2} + B_2^{\ 2}}; \tag{37}$$

$$L_1 = 1 - e^{-\xi C_1} \cos(-\xi D_1); \tag{38}$$

$$M_1 = -e^{-\xi C_1}\sin(-\xi D_1). \tag{39}$$

A real part of the transfer function (19) was separated:

$$O_3(\omega) = K_f \chi_p L_{hw}(C_1 L_1) - (D_1 M_1). \tag{40}$$

To use the real part $O_3(\omega)$, the following factors were obtained:

$$A_1 = -T_m \omega^2; \tag{41}$$

$$A_2 = \varepsilon\ - T_{hw} T_m \omega^2; \tag{42}$$

$$B_1 = (\varepsilon + 1)\omega; \tag{43}$$

$$B_2 = T_{hw}\varepsilon\omega + T_{hw}\omega + T_m\omega + \varepsilon, \tag{44}$$

$$C_1 = \frac{A_1 A_2 + B_1 B_2}{A_2^2 + B_2^2}; \tag{45}$$

$$D_1 = \frac{A_2 B_1 - A_1 B_2}{A_2^2 + B_2^2}; \tag{46}$$

$$L_1 = 1 - e^{-\xi C_1}\cos(-\xi D_1); \tag{47}$$

$$M_1 = -e^{-\xi C_1}\sin(-\xi D_1). \tag{48}$$

The transfer functions (17–19), which were obtained based on the use of the operator method of solving the system of nonlinear differential equations, retains the Laplace transform parameter – S ($S = \omega j$), where ω is the frequency, 1/s. To switch from the frequency area to the time area, a real part ((20), (31), (40)), obtained as a result of the mathematical treatment of transfer functions, was separated. It is these parts that are included in the integrals (49, 50, 51), which makes it possible to obtain dynamic characteristics of a change the temperatures of heating water, refrigerant consumption, the number of revolutions of the compressor motor, respectively using the inverse Fourier transform:

$$t_{hw}(\tau, z) = \frac{1}{2\pi}\int_0^\infty O_1(\omega)\sin(\tau\omega/\omega)d\omega. \tag{49}$$

$$G_r(\tau, z) = \frac{1}{2\pi}\int_0^\infty O_2(\omega)\sin(\tau\omega/\omega)d\omega. \tag{50}$$

$$n(\tau) = G_r(\tau, z)K_f(\tau) = \frac{1}{2\pi}\int_0^\infty O_3(\omega)\sin(\tau\omega/\omega)d\omega. \tag{51}$$

where t_{hw} – the temperature of heating water, K; G_r – refrigerant consumption, kg/s; n – the number of revolutions of the compressor motor, rpm

So, for obtaining the reference estimation of heating water temperature change, refrigerant consumption, number of revolutions of the compressor motor a block diagram (Figure. 2.5) is proposed using, for example, the initial data of constructive-regime implementation of a methane tank producing 352.5 m^3/day of biogas.

The biogas plant with a capacity of 325 m^3/day of biogas production includes the following components: the heat exchanger fitted in the methane tank with wort consumption 0.4 kg/s., heating water consumption – 0.145 kg/s. The heat pump system type, Vailliant VWW 41/3 (Germany) with a heating capacity of 19.8 kW, consumed electric power 3.5 kWm, *COP* – coefficient of performance of the heat pump system 5.7. Boundary change of refrigerant temperature at the inlet of condencer: 60…50°C, boundary change of the heating water at the outlet of the condenser 55…45°C and temperature of the wort at the outlet of the evaporator: 21…4°C.

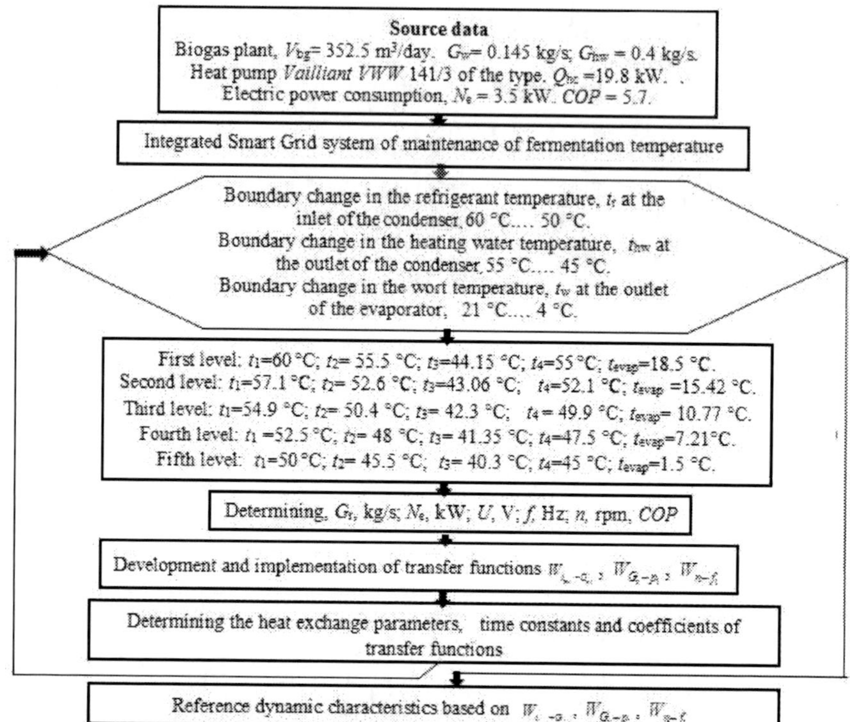

Figure 2.5. Block diagram of complex mathematical modeling of the heat pump power supply of the biogas plant: V_{bg} – productivity of the biogas plant, m³/day; G_w, G_{hw} – wort consumption, heating water consumption, respectively, kg/s; t_1, t_2 – the temperature of the refrigerant at the inlet to the condenser and at the outlet of the condenser, respectively, °C; t_3, t_4 – the temperature of the heated water at the inlet to the condenser and at the outlet of the condenser, respectively, °C; *U* – voltage, V; *f* – voltage frequency, Hz; t_r, t_{hw}, t_w, t_{evap}, – the temperature of the refrigerant, heating water, wort, evaporation of the refrigerant, respectively, °C; *n*– the number of revolutions of the compressor electric motor, rpm; *COP* – coefficient of performance of the heat pump system.

The following levels of operation of the heat pump system have been established for the change in the temperature of the refrigerant at the inlet to the condenser and at the outlet from the condenser: – first level: 60–55.5°C; second level: 57.1–52.6°C; third level: 54.9–50.4°C; fourth level: 52.5–48°C; fifth level: 50°C –45.5°C. They correspond to changes in the temperature of heating water: 44.15–55°C; 43.06–52.1°C; 42.3–49.9°C; 41.35–47.5°C. 40.3–45°C and refrigerant evaporation temperature 18.5°C, 15.42°C, 10.77°C, 7.21°C, 1.5°C.

Table 2.4. Operating parameters of the heat pump system

Levels of operation	G_r, kg/s	N_e, kW	U, V	f, Hz	n,. rpm	COP
First level	0.1196	3.14	400	50	1500	5.79
Second level	0. 1026	2.71	345.2	43.5	1294.5	5.59
Third level	0.0806	2.30	293	36.60	1098	5.53
Fourth level	0. 0641	1.91	243.3	30.41	912.3	5.38
Fifth level	0. 0480	1.58	201.3	25.17	755.1	5.00

Note: Gr – refrigerant consumption, kg/s; Ne – power of the compressor electric motor, kW; U – voltage, V; f – voltage frequency, Hz; n – the number of revolutions of the compressor electric motor, rpm; COP – coefficient of performance of the heat pump system.

Table 2.5. Heat transfer parameters as part of complex mathematical modeling of heat pump power supply

Levels of operation	Parameter		
	α_r, kW/(m^2·K)	α_{hw}, kW/(m^2·K)	k, kW/(m^2·K)
First level	1.569	1.475	0.743
Second level	1.681	1.589	0.796
Third level	1.808	1.653	0.841
Fourth level	2.054	1.727	0.911
Fifth level	2.530	1.828	1.033

Note: αr – coefficient of convective heat transfer from the refrigerant to the condenser wall, kW/(m2·K); αw – coefficient of convective heat transfer from the condenser wall to heating water, kW/(m2·K); k – heat transfer coefficient, kW/(m2·K).

Table 2.6. Time constants and coefficients of the mathematical model of the dynamics of the heat pump power supply

Levels of operation	T_{hw}, s	T_m, s	L_{hw}., m	ε	L_r., m	L_r^*	ε^*	ζ
First level	3.20	4.10	75.39	1.2635	7.40	0.1190	0.1131	0.4097
Second level	2.97	3.81	69.98	1.2760	5.34	0.1577	1.0748	0.4320
Third level	2.86	3.66	67.28	1.2993	3.85	0.2061	1.0315	0.4415
Fourth level	2.74	3.50	64.40	1.4128	2.0	0.3333	0.9419	0.4438
Fifth level	2.58	3.31	60.82	1.6432	1.14	0.4673	0.8753	0.4340

According to formulas (1, 2)–(17 – 19), block diagram (Figure 2.5) the results of complex mathematical modeling of the heat pump power supply of the biogas plant with thermophysical properties of R 134a refrigerant are presented (Tables 2.4 – 2.6).

As presented in Table 2.6 the time constants and coefficients that are part of the mathematical model of dynamics (17–19) are obtained on the basis of parameters as part of complex mathematical modeling (Tables 2.4, 2.5).

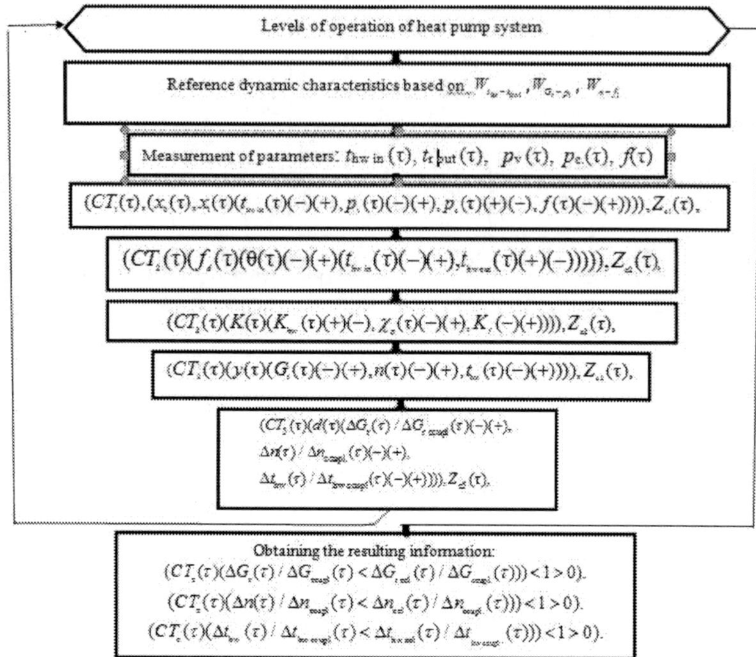

Figure 2.6. Block diagram of control of efficiency of the heat pump system: $t_{hw\,in}$, $t_{hw\,out}$, $t_{r\,out}$,, θ are the temperature of heating water at the inlet of the condenser and the outlet of the condenser, the temperature of refrigerant at the outlet of the condenser, the temperature of the separating wall, K, respectively; p_v, p_c are the evaporation pressure, condensation pressure, MPa, respectively; G_r is the refrigerant consumption, kg/s; f is the voltage frequency, Hz; f_d is the diagnosed parameters; n is the number of revolutions of the compressor electric motor, rpm; CT is the event control; Z *is* the logical relations; d is the dynamic parameters; x is the influences; y is the output parameters; K *is* the coefficients of mathematical description; τ is the time, s. Indices: c is the efficiency control; ccupl, ccl are constant, calculated value of the parameter of the upper level of functioning, level of functioning, respectively; 0, 1, 2 are the initial stationary mode, external, internal influences; 3 is the coefficients of dynamics equations; 4 is the essential predicted parameters; 5 is dynamic parameters.

According to formulas (1, 2)–(17–19) the block diagram for the control of workability of the heat pump system to support the operation of the biogas plant (Figure 2.6) is developed. The functional evaluation of changes in heating water temperature, refrigerant consumption, speed of the heat pump compressor electric motor are obtained.

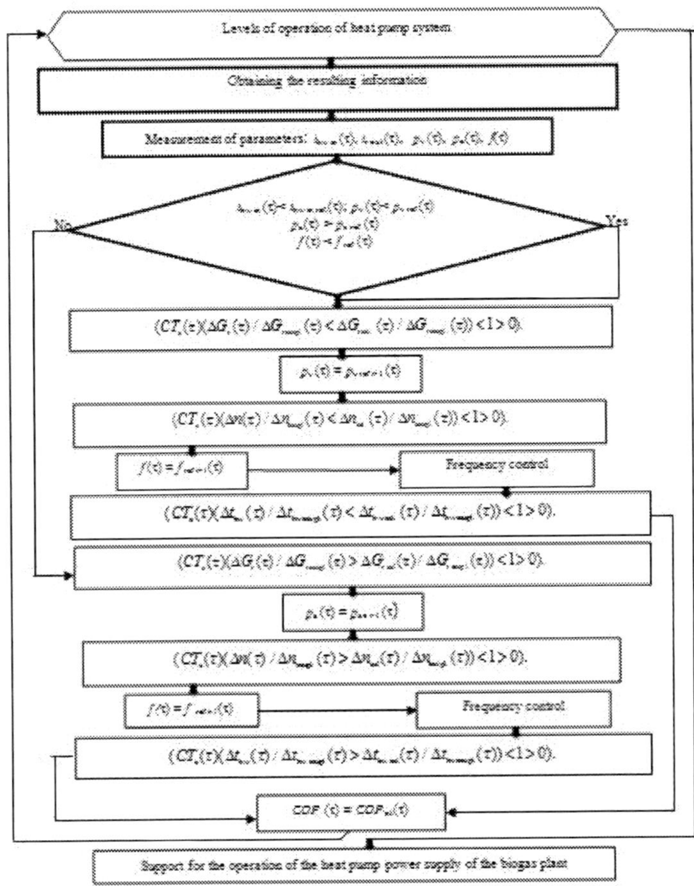

Figure 2.7. Block diagram of support for the operation of the heat pump power supply of the biogas plant: $t_{r\ out}$, t_{hw} are the temperature of refrigerant at the outlet of the condenser, heating water at the outlet of the condenser, K, respectively; p_v, p_c are the evaporation pressure, condensation pressure, MPa, respectively; G_r is the refrigerant consumption, kg/s; f is the voltage frequency, Hz; n is the number of revolutions of the compressor electric motor, rpm; *COP* is the coefficient of erformance of the heat pump system; τ is the time, s. Indices: ref is the reference value of the parameter; ccupl, ccl are constant, calculated value of the parameter of the upper level of functioning, level of functioning, respectively; nl is the new level of functioning.

According to Formulas (1, 2) - (17–19) decision-making to support the operation of heat pump power supply of biogas plant (Figure 2.7) is based on obtaining the resulting information of control of workability of the heat pump system (Figure 2.6).

A complex integrated system for supporting the functioning of heat pump power supply of biogas plant has been developed (Tables 2.7, 2.8).

Table 2.7. Integrated Smart Grid system of maintenance of heating water temperature

Time, τ, 10^3 s	Change in heating water temperature	$\Delta t_{hw}(\tau)/\Delta t_{hw}(\tau)_1$	$t_{hw}(\tau)$, °C
13	Loading fresh material. $t_{r\,in}$=60°C; $t_{r\,out}$ = 55.5°C; $t_{hw\,in}$ =44.15°C; f=50Hz	1	55
26	Discharge – charge; $t_{r\,in}$ = 60°C; $t_{r\,out}$ = 58°C; $t_{hw\,in.}$ =43.6°C; f=49.95Hz	0.6821	51.82
39	Decision making. $t_{r\,in}$ = 57.1°C; $t_{r\,out}$ = 52.6°C; $t_{hw\,in}$ =43.6°C; f=43.15Hz	0.6806	51.80
52	Discharge – charge $t_{r\,in}$ =57.1°C; $t_{r\,out}$ = 54.5°C; $t_{hw\,in}$ =42.6°C; f=40Hz	0.4491	49.48
65	Decision making. $t_{r\,in}$ =54.9°C; $t_{r\,out}$ = 50.4°C; $t_{hw\,in.}$ =42.6°C f=36.6Hz	0.4426	49.41
78	Discharge – charge $t_{r\,in}$ =54.9°C; $t_{r\,out}$ = 52.6°C; $t_{hw\,in}$ =41.5°C; f=33.5Hz	0.2417	47.40
91	Decision making $t_{r\,in}$ =52.5°C; $t_{r\,out}$ = 48°C; $t_{hw\,in}$ =41.5°C; f=30.41Hz	0.2364	47.35
104	Discharge – charge $t_{r\,in}$ =52.5°C; $t_{r\,out}$ = 50.6°C; $t_{hw\,in}$ =40°C; f=25.5Hz	0.1079	46.06
117	Decision making $t_{r\,in}$ =50°C; $t_{r\,out}$ = 45.5°C; $t_{hw\,in}$ =40.3°C; f=25.17Hz	0	45
130	$t_{r\,in}$=50°C; $t_{r\,out}$ = 49.8°C; $t_{hw\,in}$ =37.04°C. Unloading fermented wort	0	45

Note: $t_{r\,in}$, $t_{r\,out}$ are refrigerant temperatures at the inlet to the condenser, at the outlet from the condenser, respectively, °C; $t_{hw\,in}$, $t_{hw\,out}$ are heating water temperature at the inlet to the condenser, at the outlet from the condenser, respectively, °C; f – voltage frequency, Hz; τ – time, s. Indexes: hw – internal flow – heating water; 1 – constant, calculated value of the parameter of the upper level of functioning.

Table 2.8. Integrated Smart Grid system of maintenance of consumption of refrigerant vapor and the number of revolutions of the compressor electric motor

Time, τ, 10^2 s	Changing the parameters	$\Delta n(\tau)$ $/\Delta n_1(\tau)$	$\Delta G_r(\tau)$ $/\Delta G_1(\tau)$	$n(\tau)$, rpm	$G_r(\tau)$, kg/s
13	Loading fresh material t_1=60°C; t_2= 55.5°C; p_e=0.5466 MPa; p_c=1.5191 MPa; f=50 Hz	1	1	1500	0.1196
26	Discharge – Charge t_1=60°C; t_2= 58°C; p_e=0,4961 MPa; p_c=1.6090 MPa; f=49.8 Hz	0.9298	0.9298	1394.7	0.1112
39	Decision making t_1=57.1°C; t_2= 52.6°C; p_e=0.4961 MPa; p_c=1.4128 MPa; f=43.15 Hz	0.8730	0.8730	1309.5	0.1044
52	Discharge – Charge t_1=57.1°C; t_2= 54.5°C; $\boldsymbol{p_e}$**=0.4251** MPa; p_c=1.4843 MPa; f=40 Hz	0.7591	0.7591	1138.65	0.0908
65	Decision making. t_1=54.9 °C; t_2= 50.4 °C; p_e=0.4251 MPa; p_c=1.3776 MPa; f=36.6 Hz	0.7460	0.7460	1119	0.0892
78	Discharge – Charge t_1=54.9 °C; t_2= 52.6 °C; f=33.5 Hz; p_e=0.3777 MPa; p_c=1.4154 MPa	0.6444	0.6444	966.6	0.0771
91	Decision making. t_1=52,5 °C; t_2= 48 °C; p_e=0.3777 MPa; p_c=1.2543 MPa; f=30.41 Hz	0.6377	0.6377	956.55	0.0763
104	Discharge – Charge t_1=52.5 °C; t_2= 50.6 °C; p_e=0.3095 MPa; p_c=1.3428 MPa; f=25.5 Hz	0.4733	0.4733	709.95	0.0790
117	Decision making t_1=50 °C; t_2= 45.5 °C; p_e=0.3095 MPa; p_c=1.1750 MPa; f=25.17 Hz	0.4584	0.4584	687.6	0.0566
130	t_1=50°C; t_2= 49.8°C; t_{outlet} =37.04°C. Unloading fermented wort	0.4584	0.4584	687.6	0.0566

Note: t_1,t_2 –the temperatures of the refrigerant at the inlet to the condenser of the heat pump and at the outlet of the condenser of the heat pump, respectively, °C; t_{outlet}– measured temperature of heating water at the outlet of the heat exchanger fitted in the methane tank, °C; p_e – evaporation pressure, MPa; p_c – condensing pressure, MPa; f – voltage frequency, Hz; τ – time, s. Indexes: 1 – constant, calculated value of the parameter of the lower level of functioning.

The temperature of heating water from outlet of the condencer in the established period is determined as follows (Table 2.7):

$$t_{i+1}(\tau) = t_i - \\ -\begin{pmatrix} \Delta t_i(\tau)/\Delta t_1(\tau) - \\ -\Delta t_{i+1}(\tau)/\Delta t_1(\tau) \end{pmatrix}(t_1 - t_2), \tag{52}$$

where t is the temperature of heating water, °C; t_1, t_2 are the initial, final values of temperature of heating water; τ is the time, s. Index: 1 is the constant calculation value of the parameter of the upper level of operation; i is the number of levels of functioning of the biogas plant-heat pump system

The refrigerant consumption in the established period is determined as follows (Table 2.8):

$$G_{i+1}(\tau)=\left(\Delta G_{i+1}(\tau)/\Delta G_{1.}(\tau)\right)G_{in}, \quad (53)$$

where G – refrigerant consumption, kg/s; G_{in} – initial value of refrigerant consumption, kg/s; i – the number of levels of functioning; τ – time, s. Index 1 – constant, calculated value of the parameter of the upper level of functioning.

The number of revolutions of the compressor electric motor in the established period is determined as follows (Table 2.8):

$$n_{i+1}(\tau)=\left(\Delta n_{i+1}(\tau)/\Delta n_{\text{с.р.верх..}}(\tau)\right)n_1, \quad (54)$$

where n – number of revolutions of the compressor electric motor, rpm; n_1 – initial value of number of revolutions of the compressor electric motor, rpm; i – the number of levels of functioning; τ – time, s. Index 1 – constant, calculated value of the parameter of the upper level of functioning.

Thus, for example, in the time interval $65 \cdot 10^2$ s (1.8 hours) from the loading of fresh material into the biogas plant (Table 2,7, Figure 2.8), the decision was made to set the temperature of the heating water at the inlet of the heat exchanger, fitted in a methane tank, at the level of 49.41°C.

The temperature of heating water using formula (52) is:

49,41°C=49,48°C- (0,4491–0,4426)(55°C–45°C).

The refrigerant consumption, the number of revolutions of the electric motor of the compressor of the heat pump (Table 2.8, Figure 2.9) using formulas (53) (54) are:

0.0892 kg/s =0.7460(0,1196 kg/s),

1119 rpm = 0.7460(1500 rpm).

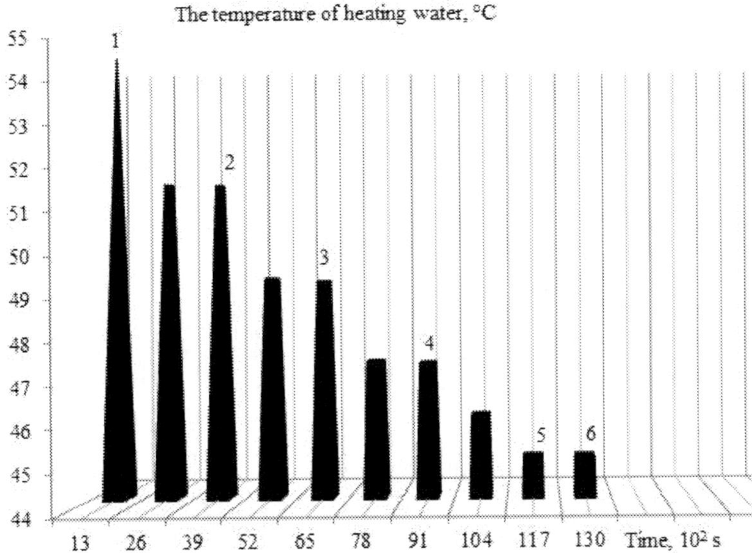

Figure 2.8. Support of the heating water temperature at the inlet of the heat exchanger fitted in the methane tank on the basis of frequency control of the electric motor of the heat pump compressor, where 2, 3, 4. 5 - decision making to change the power of the heat pump, 1, 6 - supporting the loading of fresh wort and unloading of fermented material, respectively.

The proposed systems are based on forecasting changes in heating water temperature in relation to changes in refrigerant vapor consumption, number of revolutions of the compressor electric motor. There is a continuous measurement of the temperature of the heating water at the outlet of the heat exchanger built into the methane tank, the temperature of the refrigerant at the outlet of the condenser, evaporation pressure, condensing pressure to determine their ratio and voltage frequency.

The temperature of the fermentation of raw material is maintained at the level of 35.62°C (Table 2.3).

The temperature of the fermentation of raw material is maintained at the level of 35.62°C (Table 2.3).

Based on forecasting changes in heating water temperature in relation to changes in refrigerant vapor consumption, number of revolutions of the compressor electric motor (Tables 2.3, 2.7, 2.8) developed a comprehensive integrated system to support the operation of biogas plants based on frequency control of heat pump power supply, low-potential energy source for which is fermented wort (Tables 2.9, Figure 2.10).

The developed integrated system for maintaining the fermentation temperature (Table 2.9, Figure 2.10) is the basis for a comprehensive mathematical modeling of the cogeneration system on biogas fuel.

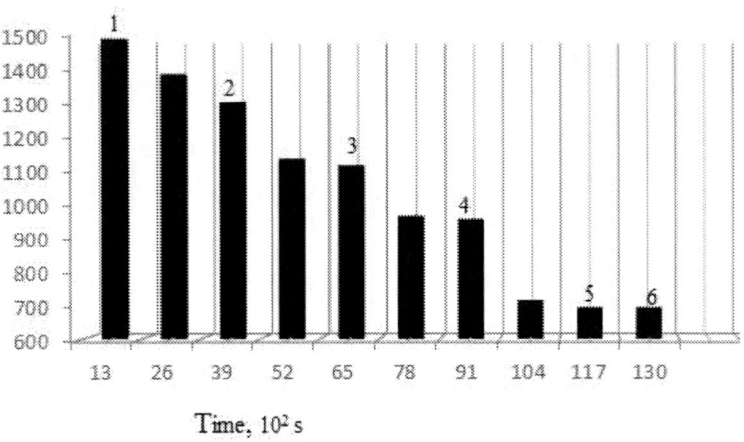

Figure 2.9. Changing the number of revolutions of the compressor electric motor based on frequency control, where 2, 3, 4. 5 - decision making to change the power of the heat pump, 1, 6 - supporting the loading of fresh wort and unloading of fermented material, respectively.

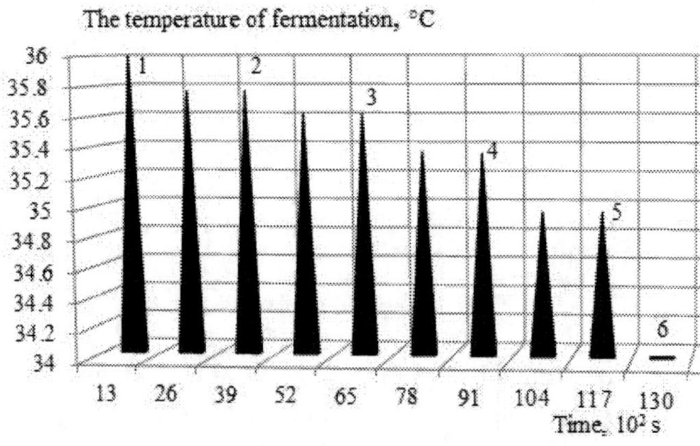

Figure 2.10. Support of the fermentation temperature based on frequency control of the heat pump compressor electric motor, where 2, 3, 4. 5 - decision-making to change the power of the heat pump, 1, 6 - support for loading fresh wort and unloading of fermented material, respectively.

Table 2.9. Integrated Smart Grid system of maintenance of fermentation temperature using heat pump power supply

Time, τ, 10^2 s	Change in fermentation temperature	$\Delta t(\tau)/\Delta t(\tau)_1$	$t(\tau)$, °C
13	Loading fresh material Setting the temperature t_{in}=55°C. N_t=3.14 kW; f=50 Hz; n=1500 rpm; G_r=0.130 kg/s; U=400 V; COP=5.79	1	36
26	Discharge – Charge t_{out}=43.6°C	0.8874	35.77
39	Decision making Setting the temperature t_{in}=52.1°C. N_t =2.71 kW; f=43.15 Hz; n=1294.5 rpm; G_r=0.102 kg/s; U=345.2 V; COP=5.59.	0.8866	35.77
52	Discharge – Charge t_{out}=42.6°C	0.8130	35.62
65	Decision making Setting the temperature t_{in}=49.9°C. N_t =2.3 kW; f=36.6 Hz; n=1098 rpm; G_r=0.08 kg/s; U=293 V; COP=5.53	0.8119	35.62
78	Discharge – Charge t_{out}=41.5°C	0.6871	35.37
91	Decision making Setting the temperature t_{in} =47.5°C. N_t =1.91 kW; f=30.41 Hz; n=912.3 rpm; G_r=0.048 kg/s; U=243.3 V; COP=5.38.	0.6823	35.36
104	Discharge – Charge t_{out}=40°C	0.4872	34.97
117	Decision making Setting the temperature t_{in}=45°C N_t =1.58 kW; f=25.17 Hz; n=755.1 rpm; G_r=0.034 kg/s; U=201.3 V; COP=5	0.4870	34.97
130	t_{out}=37.04°C. Unloading fermented wort	0	34

Note: G_r – refrigerant consumption, kg/s;, N_t – power of the compressor electric motor of the heat pump, kW; t_{in} –setting temperature of heating water at the inlet of the heat exchanger fitted in the methane tank, t_{out} – measured temperatures of heating water at the outlet of the heat exchanger fitted in the methane tank ℃ t – fermentation temperature, ℃; U – voltage, V; f – voltage frequency, Hz; τ – time, s. n – number of revolutions of the compressor electric motor, rpm; COP – коефіцієнт продуктивності теплонасосної системи. Index 1 – constant, calculated value of the parameter of the lower level of functioning.

Smart Grid System to Support the Operation of the Cogeneration System at the Decision-Making Level

According to formulas (1), (2), the prediction of a change in the power factor of the cogeneration system and the temperature of local water was proposed.

The voltage at the inlet of the inverter, at the outlet of the inverter and voltage frequency, is measured. In the engine cooling circuit, the following parameters are measured: the temperature of the cooling water at the inlet to the heat exchanger at the outlet of the heat exchanger and the return water temperature.

Transfer function by the channel "power factor of the cogeneration system – voltage at the inlet to the inverter" is complex. A change in power factor and the local water temperature are estimated. A change in the local water temperature is estimated both over time and along the spatial coordinate of the length of the plate of the heat exchanger of the engine cooling circuit:

$$W_{PF-U_1} = \frac{K_{\text{pf}} K_{\text{ecw}} \varepsilon \left(1 - L_{\text{ecw}}{}^{*}\right)}{\left(T_{\text{lw}} S + 1\right)\beta - 1}\left(1 - e^{-\gamma\xi}\right), \tag{55}$$

where

$$K_{\text{pf}} = \frac{I_2\left(U_1 - U_2\right)}{I_1 U_1};\; K_{\text{ecw}} = \frac{m\left(\theta_0 - \sigma_0\right)}{G_{\text{ecw}0}};\; \varepsilon = \frac{\alpha_{\text{ecw}0} h_{\text{ecw}0}}{\alpha_{\text{lw}0} h_{\text{lw}0}};$$

$$L_{\text{ecw}} = \frac{G_{\text{ecw}} C_{\text{ecw}}}{\alpha_{\text{ecw}0} h_{\text{ecw}0}};\; T_{\text{lw}} = \frac{g_{\text{lw}} C_{\text{lw}}}{\alpha_{\text{lw}0} h_{\text{lw}0}};\; \beta = T_{\text{m}} S + \varepsilon^{*} + 1;$$

$$T_{\text{m}} = \frac{g_{\text{m}} C_{\text{m}}}{\alpha_{\text{lw}0} h_{\text{lw}0}};\; \varepsilon^{*} = \varepsilon\left(1 - L_{\text{ecw}}{}^{*}\right);\; \gamma = \frac{\left(T_{\text{lw}} S + 1\right)\beta - 1}{\beta};$$

$$\xi = \frac{z}{L_{\text{lw}}};\; L_{\text{lw}} = \frac{G_{\text{lw}} C_{\text{lw}}}{\alpha_{\text{lw}0} h_{\text{lw}0}},$$

where PF is the power factor of the cogeneration system; I_1, I_2 are the current at the inlet of the inverter and at the outlet of the inverter, respectively, A; U_1, U_2 are the voltage at the inlet of the inverter and at the outlet of the inverter, respectively, V; C is the specific thermal capacity, kJ/(kg·K); α is the heat transfer factor, kW/(m^2·K); G is the loss of substance, kg/s; g is the specific weight of a substance, kg/m; h is the specific surface, m^2/m; σ, θ are the temperature of cooling water and of separating wall, respectively, K; z is the coordinate of the length of the heat exchanger, m; T_{lw}, T_{m} are the time constants that characterize the thermal accumulating capacity of local water, metal, s; m is the indicator of the dependence of heat transfer factor on consumption ι is the time, s; S is the Laplace parameter; $S=\omega j$; ω is the frequency, 1/s. Indices: 0 – initial stationary mode; 1 – the inlet of the cogeneration system; lw – local water; ecw– engine cooling water; m– metal wall.

Transfer function by the channel "power factor of the cogeneration system – voltage at the inlet to the inverter" was obtained based on the solution of a system of nonlinear differential equations using the Laplace transform tool. The system of differential equations includes the equation of state as the estimation of the physical model of the biogas plant – heat pump system – cogeneration system. The system of differential equations also includes the equation of energy of transmitting and receiving media – engine cooling water and local water, respectively, and the equation of thermal balance for the wall of the heat exchanger of the engine cooling circuit The equation of the energy of the receiving medium is developed with the representation of a change in local water both in time and along the spatial coordinate, which coincides with the direction of the motion flow of the medium and includes the K_{ecw} factor. The equation of the energy of the transmitting medium includes the K_{pf} coefficient, which assesses a change in power factor of the cogeneration system under conditions of maintaining the ratio of production of electric power production and heat at a change of consumption.

To use the real part $O(\omega)$, the following factors were obtained:

$$O(\omega)=\frac{(L_1 A_1)+(M_1 B_1)(1-L_{ecw}^{*})}{(A_1^2+B_1^2)}. \tag{56}$$

The K_{ecw} factor includes the temperature of the separating wall θ:

$$\theta=(\alpha_{B}(\sigma_1+\sigma_2)/2)+(A(t_1+t_2)/2)/(\alpha_{lw}+A), \tag{57}$$

where σ_1, σ_2 are the temperatures of cooling water at the inlet and at the outlet of the heat exchanger of the engine cooling circuit, K, respectively; t_1, t_2 are the temperatures of local water at the inlet and at the outlet of the heat exchanger, K, respectively; α is the heat transfer factor, kW/(m^2·K). Index: lw– local water.

$$A = 1/(\delta_{\text{m}} / \lambda_{\text{m}} + 1/\alpha_{\text{ecw}}), \tag{58}$$

where δ is the thickness of a plate of the heat exchanger, m; λ is the thermal conductivity of the metal plates of the heat exchanger, kW/(m·K). Indices: ecw– cooling water; m – metal wall of the plate of a heat exchanger.

To use the real part $O(\omega)$, the following factors were obtained:

$$A_1 = \varepsilon^* - T_{\text{lw}} T_{\text{m}} \omega^2; \tag{59}$$

$$A_2 = \varepsilon^* + 1; \tag{60}$$

$$B_1 = T_{\text{lw}} \varepsilon^* \omega + T_{\text{lw}} \omega + T_{\text{m}} \omega; \tag{61}$$

$$B_2 = T_{\text{m}} \omega; \tag{62}$$

$$C_1 = \frac{A_1 A_2 + B_1 B_2}{A_2^{\ 2} + B_2^{\ 2}}; \tag{63}$$

$$D_1 = \frac{A_2 B_1 - A_1 B_2}{A_2^{\ 2} + B_2^{\ 2}}; \tag{64}$$

$$L_1 = 1 - e^{-\zeta C_1} \cos(-\xi D_1); \tag{65}$$

$$M_1 = -e^{-\zeta C_1} \sin(-\xi D_1). \tag{66}$$

The transfer function (55), which was obtained based on the use of the operator method of solving the system of nonlinear differential equations, retains the Laplace transform parameter – S (S=ωj), where ω is the frequency, 1/s. To switch from the frequency area to the time area, a real part (56), obtained as a result of the mathematical treatment of transfer function, was separated. It is this part that is included in the integrals (67), (68), which makes it possible to obtain dynamic characteristics of a change in power factor of the cogeneration system, the temperature of local water using the inverse Fourier transform.

$$PF(\tau)=\frac{1}{2\pi}\int_{0}^{\infty}K_{pf}K_{\text{ecw}}O(\omega)\sin(\tau\omega/\omega)d\omega, \tag{67}$$

$$t(\tau,z)=\frac{1}{2\pi}\int_{0}^{\infty}K_{pf}K_{\text{ecw}}O(\omega)\sin(\tau\omega/\omega)d\omega, \tag{68}$$

where *PF* is the power factor of the cogeneration system; *t* is the temperature of local water, K.

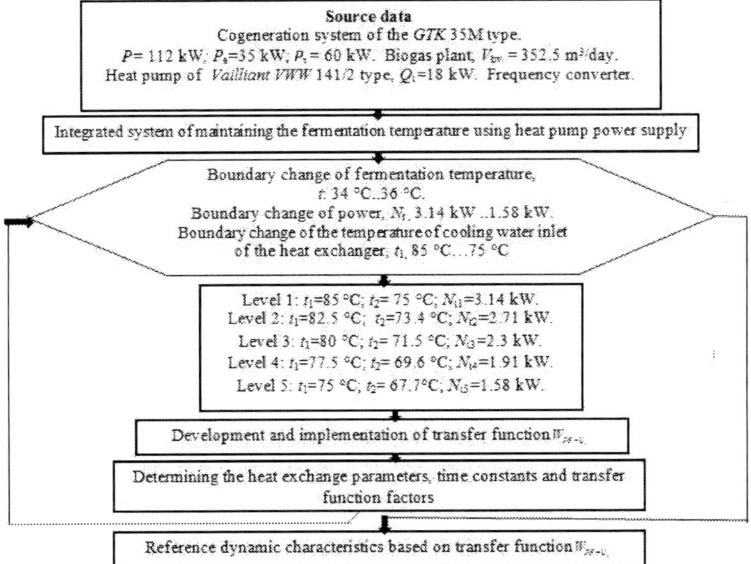

Figure 2.11. Block diagram of complex mathematical modeling of the cogeneration biogas system: *P*, P_e, P_T are the nominal, electric, thermal power of the cogeneration system, respectively, kW; Q_T, N_T are the thermal productivity, power of the thermal pump, respectively, kW; *V*bv is the biogas volume, m^3/day; *t*, t_1, t_2 are the fermentation temperature, the temperature of cooling water at the inlet of the heat exchanger of the engine cooling circuit and at the outlet of the heat exchanger, respectively, °C.

So, for obtaining the reference estimation of change in the power factor of the cogeneration system and temperature of heated local water a block diagram (Figure. 2.11) is proposed using, for example, the initial data of the complex cogeneration biogas system type *GTK* 35M with constructive-regime implementation of a methane tank producing 352.5 m^3/day of biogas and heat pump system type, *Vailliant VWW* 41/3 (Germany) with a heating capacity of 19.8 kW.

The cogeneration biogas system type *GTK* 35M. P, P_e, P_T are the nominal, electric, thermal power of the cogeneration system, 112, 35, 60, respectively, kW. Biogas plant with a capacity of 325 m^3/day of biogas production. The heat pump system type, *Vailliant VWW* 41/3 (Germany) with a heating capacity of 19.8 kW, consumed electric power 3.5 kW, *COP* – coefficient of performance of the heat pump system is 5.7.

The following levels of operation of the cogeneration system have been established for the change in the temperature of the engine cooling water at the inlet to the heat exchanger and at the outlet from the heat exchanger: – first level: 85–75°C; second level: 82.5–73.4°C; third level: 80.0–71.5°C; fourth level: 77.5–69.6°C; fifth level: 75°C –67.7°C. They correspond to changes in the power of the compressor electric motor of the heat pump system 3.14, 2.71, 2.3, 1.91, 1.58 respectively, kW.

According to formulas (1, 2)–(55), block diagram (Figure 2.11) the results of complex mathematical modeling of the cogeneration biogas system are presented (Tables 2.10, 2.11).

Table 2.10. Heat exchange parameters in the engine cooling circuit

Levels of functioning	Parameter		
	α_{ecw}, kW/(m^2·K)	α_{lw} kW/(m^2·K)	k, kW/(m^2·K)
first level	7.876	3.718	2.341
second level	7.240	3.241	2.093
third level	6.688	2.888	1.897
fourth level	6.325	2.614	1749
fifth level	6.010	2.497	1.626

Note: α_{ecw} is the factor of heat transfer from the engine cooling water to the wall of the plate of heat exchanger, kW/(m^2·K); $\alpha_в$ is the factor of heat transfer from the heat exchanger wall to local water, kW/(m^2·K); k is the heat transfer factor, kW/(m^2·K).

Table 2.11. Time constants and coefficients of mathematical model of dynamics of engine cooling circuit

Levels of functioning	T_{lw}, s	T_m, s	ε	ε^*	ζ	L_{lw}, м	L_{ecw}, m	L^*_3
first level	1. 31	0.46	2.56	2.50	0.88	43.70	42.47	0.023
second level	1. 50	0.53	2.70	2.65	0.86	50.13	48.78	0.020
third level	1. 68	0.59	2.80	2.75	0.85	56.27	54.56	0.017
fourth level	1. 86	0.65	2.90	2.88	0.83	62.16	60.28	0.016
fifth level	2. 03	0.71	3.00	2.99	0.82	67.81	65.38	0.015

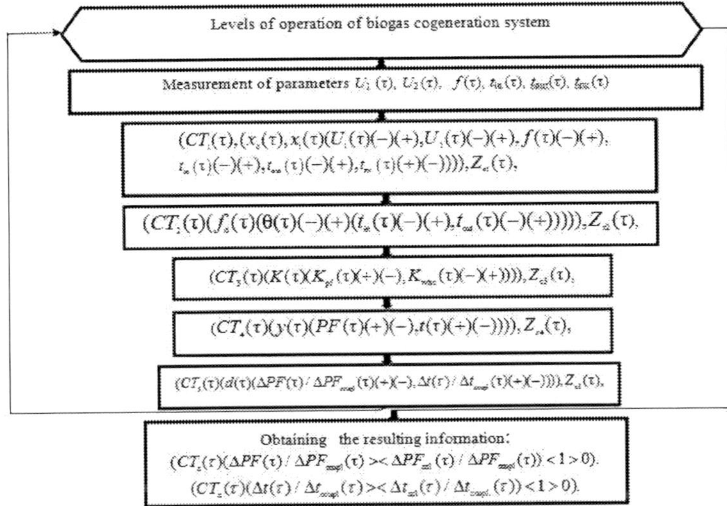

Figure 2.12. Block diagram of control of efficiency of the biogas cogeneration system: U_1, U_2 are the voltage at the inlet of the inverter and at the outlet of the inverter, respectively, V; *f* is the voltage frequency, Hz; *t*, $t_{in.}$ $t_{out,}$ $t_{rw.}$ are the temperatures of local water, cooling water at the inlet of the heat exchanger of the engine cooling circuit, at the outlet of the heat exchanger, the temperature of return water, respectively, °C; θ is the temperature of the separating wall. respectively, K; *PF* is the power factor of the cogeneration system; *CT* is the event control; *Z* is the logical relations; *d* is the dynamic parameters; *x* is the influences; f_d is the diagnosed parameters; *y* is the output parameters; *K* is the mathematic description factors; ι is time. Indices: *c* is efficiency control; ccupl is the constant calculation value of a parameter of the upper level of functioning; ccl is the constant calculation value of a parameter of the level of functioning; 0, 1, 2 are the initial stationary mode, external, internal parameters; 3 is the coefficients of dynamics equations; 4 is the essential diagnosed parameters; 5 is dynamic parameters.

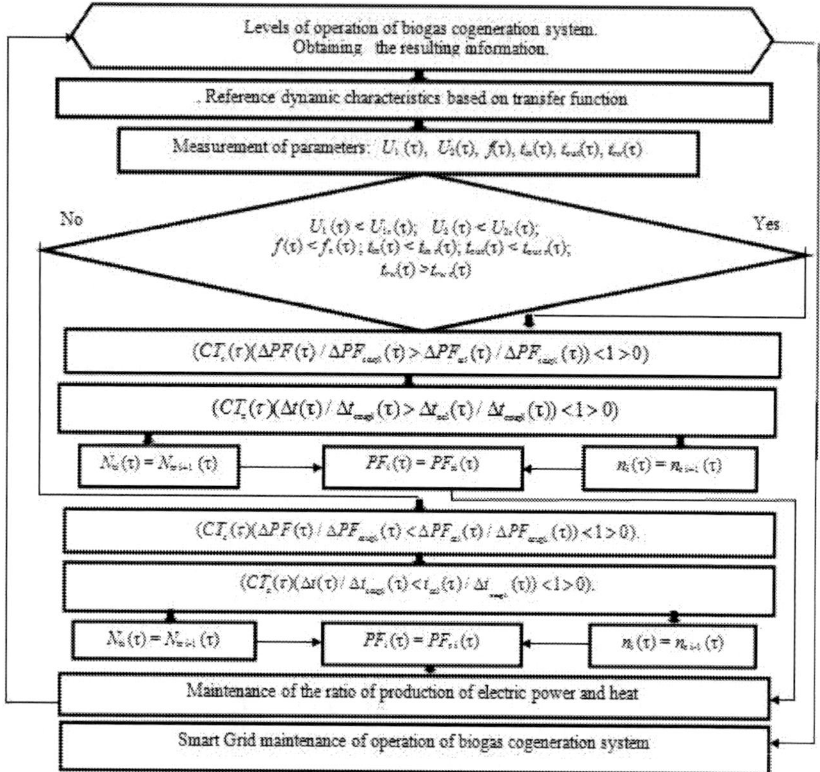

Figure 2.13. Block diagram of maintenance of functioning of the cogeneration system: U_1, U_2 are the voltages at the inlet of the invertor and at the outlet of the invertor, respectively, V; f is the voltage frequency, Hz; CT_c is the efficiency control; t, t_{in}. t_{out}, t_{rw} are the temperature of local water, cooling water at the inlet of the heat exchanger of the engine cooling circuit, at the outlet of the heat exchanger, the temperature of return water, respectively, K; PF is the power factor of the cogeneration system; N_T is the power of the heat pump, kW; n is the number of plates of the heat exchanger of the engine cooling circuit; ι is the time, Indices: i is the number of operation levels; r is the reference value of the parameter; ccupl is the constant calculation value of the parameter of the upper level of functioning; ccl is the constant calculation value of a parameter of the level of functioning.

The time constants and coefficients, which are a part of the mathematical model of dynamics (55), presented in Table 2.11, were obtained based on the integrated system for maintaining the fermentation temperature (Table 2.9, Figure 2.10) and parameters of heat exchange in the engine cooling circuit (Table 2.10).

Table 2.12. Integrated Smart Grid System of harmonization of production and consumption of electric power and heat at a change of consumption

Time, τ, 100 s	Change of parameters	$\Delta PF(\tau)/\Delta PF(\tau)_1$	$PF(\tau)$	tgφ(τ)	$Q(\tau)$, kVAR	$t(\tau)$, °C
13	Loading of fresh material N_t 3.14 kW; U_1=400 V; U_2=380 V; t_{inlet} =85°C; t_{outlet} =75°C t_{rw} =30°C; n=36 pieces; G 1.43 kg/s; P_e 30.2 kW; P_t=51.9 kW; m=0.58	0.1442	0.8644	0.5816	17.56	33.6
26	Charge – discharge N_t =2.9 kW; U_1=390 V; U_2=370 V; t_{inlet} =82.5°C; t_{outlet} =73.4,°C; t_{rw} =35°C; n=36 pieces; G=1.43 kg/s; P_e=30.2 kW; P_t =51.8 kW; m=0.58	0.1293	0.8629	0.5856	17.68	33.2
39	Decision making N_t =2.71 kW; U_1=400 V; U_2=345.2 V Decision making n=44 pieces; G=1.51 kg/s; t_{inlet} =82.5°C t_{outlet} =73.4°C; t_{rw} =30°C; P_e=31.03 kW; P_t =53.2 kW; m=0.58	0.3653	0.8865	0.5219	16.33	39.1
52	Charge – discharge N_t 2.51 kW; U_1=380 V; U_2=320 V; t_{inlet} =80°C; t_{outlet} =71.5°C;. t_{rw} =35°C; n=44 pieces; G=1.51 kg/s; P_e=31 kW; P_t =53.1 kW; m=0.58	0.3588	0.8858	0.5239	16.24	38.94
65	Decision making N_t 2.3 kW; U_1=400 V; U_2=293 V Decision making n=52 pieces. G=1.56 kg/s; t_{inlet} =80°C; t_{outlet} =71.5°C;. t_{rw} =30°C; P_e=32 kW; P_t =54.9 kW; m=0.58	0.6562	0.9155	0.4392	14.05	46.37

Table 2.12. (Continued)

Time, τ, 100 s	Change of parameters	$\Delta PF(\tau)/\Delta PF(\tau)_1$	$PF(\tau)$	tgφ(τ)	$Q(\tau)$, kVAR	$t(\tau)$, °C
78	Charge – discharge N_t=2.2 kW; U_1=390 V; U_2=280 V; t_{inlet}=77.5°C; t_{outlet}=69.6°C; t_{rw}=35°C; n=52 pieces G=1.56 kg/s; P_e=31.8 kW; P_t=54.5 kW; m=0.58	0.5830	0.9082	0.4609	14.66	44.54
91	Decision making N_t 1.91 kW; U_1=400 V; U_2=243.8 V Decision making n=60 pieces; G=1.63 kg/s; t_{inlet}=77.5°C; t_{outlet}=69.6°C; t_{rw}=30°C; P_e=32.8 kW; P_t=56.2 kW; m=0.58	0.8735	0.9372	0.3626	11.89	51.8
104	Charge – discharge N_t=1.73 kW; U_1=390 V; U_2=220 V; t_{inlet}=75°C; t_{outlet}=67.7°C; t_{rw}=35°C; n=60 pieces; G=1.63 kg/s; P_e=32.6 kW; P_t=55.8 kW; m=0.58	0.8087	0.9307	0.3931	12.81	50.18
117	Decision making N_t=1.58 kW; U_1=400 V; U_2=201.3 V; Decision making n=68 pieces; G=1.68 kg/s; t_{inlet}=75°C; t_{outlet}=67.7°C; t_{rw}=30°C; P_e=33.2 kW; P_t=57 kW; m=0.58	1	0.9498	0.3286	10.91	55
130	Unloading of fermented wort N_t=1.58 kW; U_1=400 V; U_2=201.3 V; n=68 pieces; G=1.68 kg/s; t_{inlet}=75°C; t_{outlet}= 67.7°C; t_{rw}=30°C; P_e=33.2 kW; P_t=57 kW; m=0.58	1	0.9498	0.3286	10.9	55

Note: PF is the power factor of the cogeneration system; tgφ is the factor of reactive power of the cogeneration system; Q is the reactive power of the cogeneration system, kVAR; P_e, P_t, are the active electric, thermal power of the cogeneration system, kW, respectively; N_t is the power of the heat pump, kW; U_1, U_2, are the voltage at the inlet and at the outlet of the invertor, V, respectively; t_{inlet}, t_{outlet}, t_{rw} are the temperatures of cooling water at the inlet of the heat exchanger of the engine cooling circuit, at the outlet of the heat exchanger, the temperature of return water, respectively, °C, respectively; t is the temperature of local water, °C; G is the consumption of cooling water, kg/s; n is the number of plates of the heat exchanger of the engine cooling circuit; m is the ratio of production and consumption of electric power and heat. Index: 1 – established calculation value of the parameter of the first operation level.

Based on the proposed mathematical substantiation Smart Grid maintenance of functioning of the cogeneration system (1, 2) to (55), the block diagram for the control of workability of the biogas cogeneration system (Figure 2.12) are presented.

Control of workability of the cogeneration system (Figure 2.12) enables obtaining the resulting information on decision-making about the maintenance of the functioning of the cogeneration system.

Smart Grid System of Maintaining the Operation of the Cogeneration System at the Decision-Making Level

Based on the proposed mathematical substantiation ((1), (2), (55)), the block diagram of the maintenance of the functioning of the cogeneration biogas system (Figure 2.13) is presented.

A complex integrated system for supporting the functioning of the cogeneration biogas system is developed. There is a continuous measurement of voltage at the inlet of the inverter, at the outlet of the inverter and voltage frequency. In the engine cooling circuit, the temperature of cooling water at the inlet of the heat exchanger, at the outlet of the heat exchanger, and the temperature of return water are measured.

The maintenance of the temperature of local water in the engine cooling circuit is based on the harmonization of the ratio of production of electrical power and heat at a change in consumption (Table 2.12).

Coordination of Electricity and Heat Production Based on Forecasting Changes in Power Factor and Local Water Temperature with Changes in Consumption

A power factor of the cogeneration system in the established period is determined as follows (Table 2.12, Figure 2.13):

$$PF_{i+1}(\tau) = PF_i + \\ + \begin{pmatrix} \Delta PF_{i+1}(\tau) / \Delta PF_1(\tau) - \\ -\Delta PF_i(\tau) / \Delta PF_1(\tau) \end{pmatrix} (PF_2 - PF_1), \tag{69}$$

where PF is the power factor of the cogeneration system; PF_1, PF_2 are the initial, final values of power factor; τ is the time, s. Index: 1 is the constant calculation value of the parameter of the upper level of functioning; i is the number of levels of the cogeneration system operation.

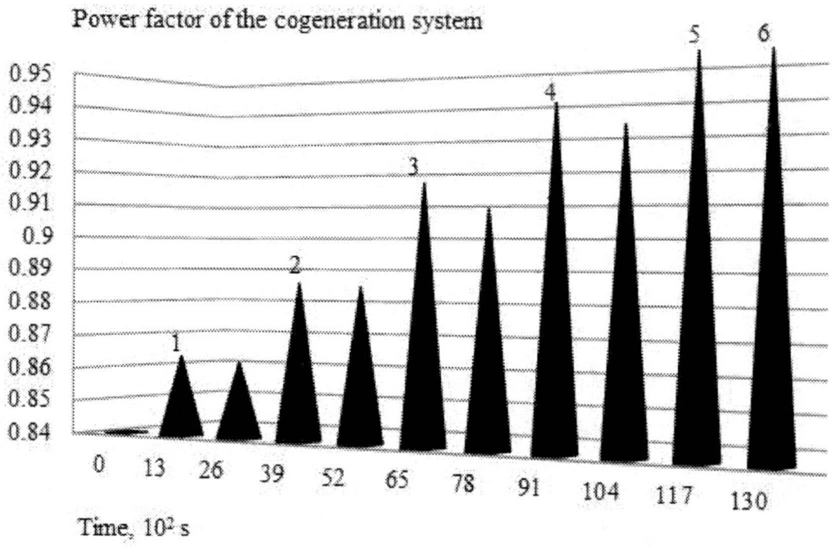

Figure 2.14. Maintenance of a change in power factor of the cogeneration system, where 2, 3, 4, 5 – making a decision on a change in power of the heat pump, 1, 6 are the maintenance of loading fresh wort and unloading of fermented material.

The temperature of local water in the established period is determined as follows (Table 2.12, Figure 2.14):

$$t_{i+1}(\tau) = t_i + \\ + \begin{pmatrix} \Delta t_{i+1}(\tau)/\Delta t_1(\tau) - \\ -\Delta t_i(\tau)/\Delta t_1(\tau) \end{pmatrix} (t_2 - t_1), \qquad (70)$$

where t is the temperature of local water, °C; t_1, t_2 are the initial, final values of temperature of local water; τ is the time, s. Index: 1 is the constant calculation value of the parameter of the upper level of operation; i is the number of levels of functioning of the cogeneration system.

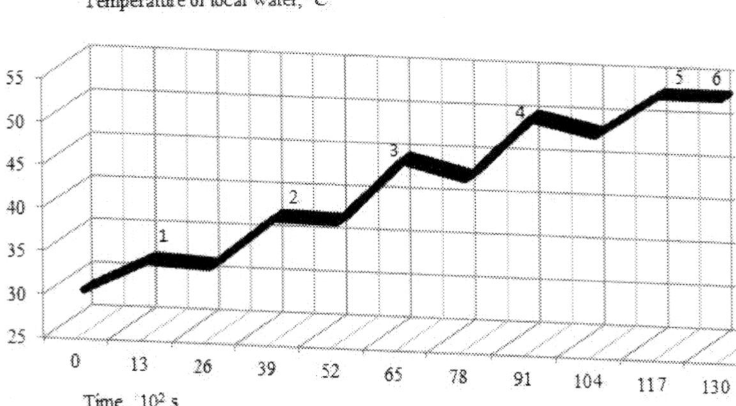

Figure 2.15. Maintenance of a change in temperature of local water in the engine cooling circuit of the cogeneration system, where 2, 3, 4. 5 are making decisions on a change of the number of plates of the heat exchanger of the engine cooling circuit, 1, 6 are the maintenance of loading of fresh wort and unloading of fermented material.

Based on the ratio $\cos\varphi = 1/\sqrt{(1+tg^2\varphi)}$ the factor of reactive power $tg\varphi$, which makes up for the period of $65 \cdot 10^2$ s (1.8 hours) 0.4392, was determined (Table 2.12, Figure 2.15). Based on the values of $\cos\varphi$ and $tg\varphi$ in this period, the active and reactive powers of the cogeneration system were found: 32 kW and 14.05 kVAR, respectively (Table 2.12, Figure 2.16).

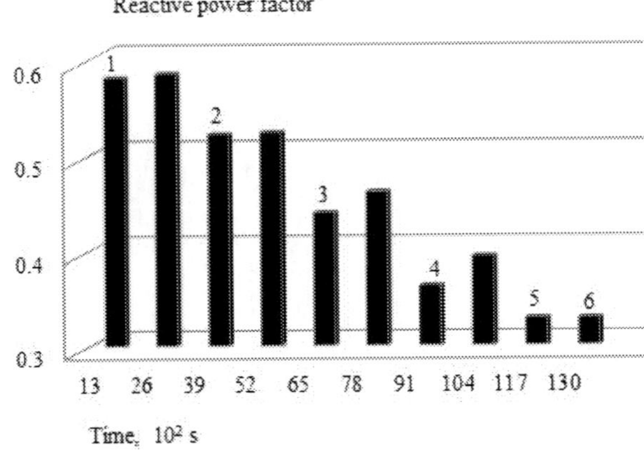

Figure 2.16. Maintenance of a change in the factor of reactive power of the cogeneration system, where 2, 3, 4, 5 are making decisions on a change in power of the heat pump, 1, 6 are maintenance of loading of fresh wort and unloading of fermented material.

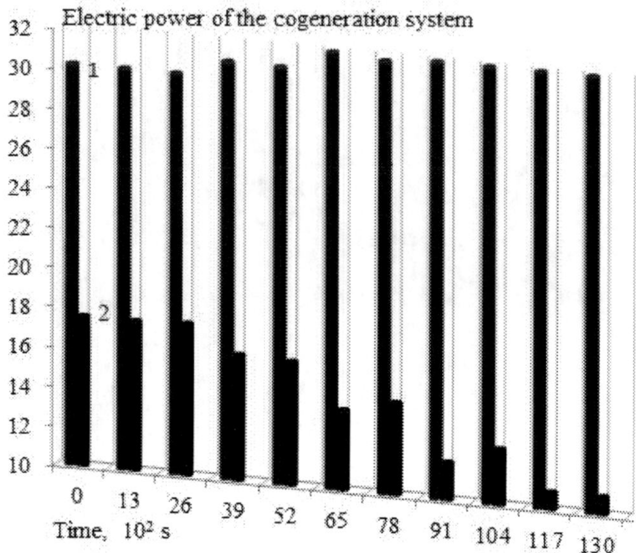

Figure 2.17. Maintenance of electric power of the cogeneration system, where 1, 2 are active power, kW, reactive power, kVAR, respectively.

Thus, for example, in the time interval $65 \cdot 10^2$ s (1.8 hours) from the loading of fresh material into the biogas plant (Table 2.9), the absolute value of power factor of the cogeneration system (Table 2.12) using formula (69) equals:

$$0{,}9155=0{,}8858+(0{,}6562-0{,}3588)(0{,}95-0{,}85).$$

The absolute value of the temperature of local water (Table 2.12) with the use of formula (70) in the period $65 \cdot 10^2$ s (1.8 hours) is:

$$46{,}37°C=38{,}94°C+(0{,}6562-0{,}3588)(55-30°C).$$

Thus, for example, in the period of time $65 \cdot 10^2$ s (1.8 hours), an increase in the power factor of the cogeneration system $PF(\tau)$ was predicted, from 0.8858 to 0.9155 (Table 2.12, Figure 2.13) by 13% relative to a decrease in the factor of reactive power tgφ from 0.5239 to 0.4392 (Table 2.12, Figure 2.15). With this aim, the forestalling decision to decrease the number of revolutions of the electric engine of the heat pump compressor to the level of 1,098 rpm (Table 2.9). A decrease in the active power of the heat pump to the level of 2.3 kW (Table 1) makes it possible to set the temperature of the warming heat carrier at the inlet of the heat exchanger, fitted in a methane tank, at the level of 49.9°C. The temperature of the fermentation of raw material is maintained at the level of 35.62°C (Table 2.9, Figure 2.10).

Predicting the maintenance of heating local water to the level of 46.37 °C in this time period (Table 2.12, Figure 2.14), the forestalling decision to increase the number of plates of the heat exchanger of the engine cooling circuit from 48 pieces to 60 pieces was made. The ratio of production of electric power and heat at a change in consumption is maintained (Table 2.12, Figure 2.17).

An increase in active electric and thermal power of the cogeneration system occurs at the compensation of reactive power of the cogeneration system up to 13% (Table 2.12, Figure 11). The implementation of such actions makes it possible to maintain the balance of generation and consumption of electrical power and heat (Table 2.12, Figure 2.17).

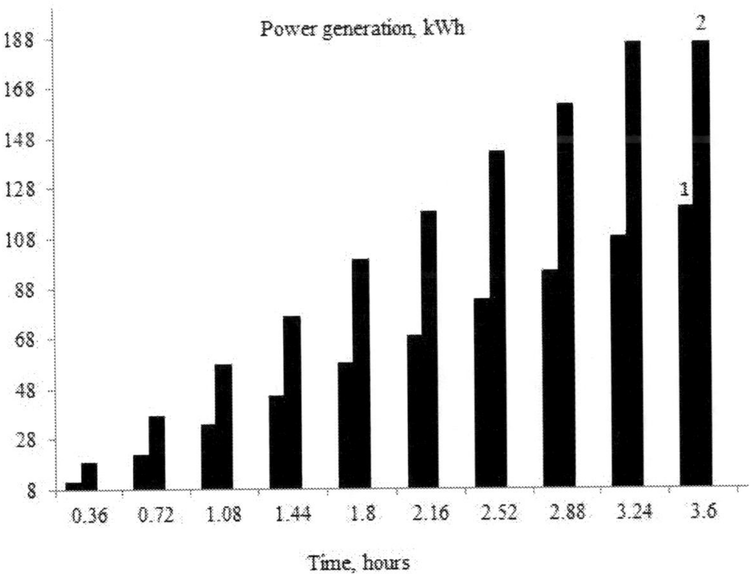

Figure 2.18. Harmonization of the generation and consumption of electric power and heat, where 1, 2 are the electric and thermal generation, respectively.

Voltage regulation in the distribution system based on a change in the reactive and active power of heat pump power supply of the biogas plant

(Table 2.12) makes it possible to maintain the ratio of production of electric power and heat at a change in consumption.

Discussion of Results of Studying the Smart Grid Technology for Maintaining the Functioning of the Cogeneration Biogas System

One of the advantages of cogeneration technologies when it comes to distributed generation of electric power is the possibility of using biogas as a renewable energy source. Optimization of the distributed generation of electric power traditionally uses the improvement of intelligent control systems without taking into account the interaction of production of electric power and consumption. The use of reactive power to control the voltage in the electric network with the help of, for example, transformers, capacitor batteries, voltage regulators, static synchronous compensators, etc. is traditional. The costs of mounting and maintaining these devices can be quite high and some of them have a relatively slow response time, of the order of a few seconds.

The inclusion of the biogas plant and the heat pump into the integrated dynamic subsystem (Figure 2.1) of the cogeneration system Smart offers an opportunity to use the heat pump as a voltage regulator in the distribution network. The changes in the reactive power of the cogeneration system are estimated based on the assessment of a change in revolutions of the electric engine of the heat pump compressor within the term of the biogas production. This term is determined by loading fresh wort and unloading fermented material, which takes place 4 –6 times per day during the continuous production of biogas. That is why it was proposed to predict a change in power factor of the cogeneration system, the temperature of local water in the engine cooling circuit based on the estimation of change in the ratio of the measured voltage at the inlet of the inverter, at the outlet of the inverter, when measuring the voltage frequency.

With this aim, the integrated Smart Grid system of maintenance of the fermentation temperature with the use of the heat pump power supply, which uses fermented wort as a low-potential energy source (Table 2.9, Figure 2.10), was developed. The developed system makes it possible while using the frequency control of the electric engine of the spiral compressor of the heat pump, to ensure a change in the temperature of the warming heat carrier at the inlet of the heat exchanger, fitted in the methane tank. It is this system that becomes the basis of complex mathematical simulation of the cogeneration system (Figure 2.11, Tables 2.10, 2.11), which results in obtaining reference information on estimation of power factor of the cogeneration system and temperature of local water. Moreover, the use of heat pumping power supply of the biogas plant makes it possible to regulate the ratio of production of electric power and heat at a change of consumption. That is why it is proposed to measure the temperature of cooling water at the inlet of the heat exchanger of the engine cooling circuit and at the outlet of the heat exchanger and the temperature of return water.

The engine cooling circuit, which usually performs the function of the protective element, becomes a comprehensive information system for assessing a change in electric and thermal power of the cogeneration system at a change in the ratio of consumption of electrical power and heat. This circuit reacts to a change in electric power consumption, regarding the frequency control of the heat pumping power supply of the biogas plant due to a change in the balance of consumption of electric power and heat. That is why using mathematical substantiation of the architecture of the cogeneration system (Figure 2.1), maintenance of functioning of the cogeneration system (Figure 2.1), and transfer function (55), it was proposed to make forestalling decisions to change the power of the heat pumping power supply of the biogas plant and to change the number of plates of the heat exchanger of the engine cooling circuit.

Voltage in the distribution system and the ratio of production of electric power and heat are maintained at a change in consumption. For this purpose, a functional estimation of a change in the power factor of the cogeneration system, the temperature of local water in the circuit of cooling the heated engine were obtained (Figure 2,12). The prediction of a change in power factor of the cogeneration system and the temperature of heated local water makes it possible to take forestalling decisions to change the power of a heat pump and to change the number of plates of the heat exchanger of the engine cooling circuit. For this purpose, the integrated Smart Grid System of harmonization of production and consumption of electric power and heat at a change of consumption (Table 2.12) was obtained, based on logical modeling (Figure 2.13), It is this integrated system that determines the exact term of making forestalling decisions, regarding the maintenance of functioning of the cogeneration biogas system.

Thus, in the period from the loading of fresh material to unloading of fermented wort (3.6 hours), the increase in power coefficient of the cogeneration system $PF(\tau)$ from 0.8644 to 0.9498 (Table 2.12, Figure 2.14) regarding an increase in reactive power factor tgφ from 0.5816 to 0.3286 (Table 2.12, Figure 2.16) was predicted. With this aim, a proactive decision was made to reduce the number of revolutions of the electric engine of the heat pump compressor to the level of 755.1 rpm (Table 2.9). A decrease in the active power of the heat pump to the level of 1.58 kW (Table 2.9) makes it possible to set the temperature of the warming heat carrier at the inlet of the heat exchanger, fitted in the methane tank, at the level of 45°C. The temperature of the fermentation of raw materials is maintained at the level of 34°C (Table 2.9, Figure 2.10).

Predicting the maintenance of heating local water to the level of 55°C in the time period (Table 2.12, Figure 2.14), the forestalling decision was made to increase the number of plates of the heat exchanger of the engine cooling circuit up to 68 pieces. The ratio of production of electric power and heat at a change in consumption is maintained (Table 2.12, Figure 2.18). An increase in active electric and thermal power of the cogeneration system was estimated by the compensation of reactive power of the cogeneration system from 17.56 kVAR to 10.9 kVAR – up to 40% (Table 2.12, Figure 2.17). The implementation of such actions makes it possible to maintain the balance of generation and consumption of electric power and heat (Table 2.12, Figure 2.18). Maintaining the operation of the cogeneration biogas system with the use of the developed Smart Grid technology allows decreasing the cost of production of electric power and heat up to 30%.

Chapter 3

Smart Grid Technology for Maintaining the Functioning of a Biodiesel Cogeneration System

Abstract

The integrated Smart Grid System of harmonization of production and consumption of electric power and heat is developed. The integrated dynamic subsystem of the biodiesel cogeneration system includes the following components: the electric network, the cogeneration unit, the engine cooling circuit as a part of the cogeneration plant, the biodiesel plant, the heat exchanger for heating oil with biodiesel heat. A change in the ratio of production and consumption of electric power and heat, change in the temperature of oil and change in the temperature of local water of the engine cooling circuit is predicted by measuring temperature of biodiesel at the outlet from the heat exchanger, the temperature of cooling water at the outlet of the heat exchanger the engine cooling circuit and the temperature of return water.

Making forestalling decisions to support the supply of fresh oil for heating, to support the supply of heated oil for etherification, changing number of plates of the biodiesel heat exchanger and the changing number of plates of the heat exchanger of the engine cooling circuit makes it possible to maintain the ratio of production and consumption of electric power and heat and the temperature of the heated oil and local water. The complex mathematical and logical modeling of the cogeneration system, based on the mathematical substantiation of the architecture of the cogeneration system and mathematical substantiation of the maintenance of functioning of the cogeneration system, was performed.

Time constants and coefficients of the mathematical models of dynamics regarding the estimation of a change in the temperature of oil, temperature of local water, were determined. Functional estimation of a change in the ratio of production and consumption of electric power and heat of the cogeneration system in the range of 0.6606 – 0.6611, temperature of oil in the range of 20–45°C, temperature of local water in the range of 30–55°C was obtained. Determining final functional information provides an opportunity to make forestalling decisions to changing the ratio of production and consumption of electric power and heat to maintain the functioning of the cogeneration system.

Keywords: Smart Grid technologies, cogeneration system, biodiesel plant, the heat exchanger for heating oil with biodiesel heat

Introduction

Biodiesel is a methyl ester produced by the esterification process of any heated vegetable oil or animal fat. To do this, one mass unit of methanol is added to seven mass units of the raw material using a catalyst. The resulting by-product, glycerin, is commonly used in the production of technical detergents. In the production of biodiesel, both traditional and the latest technologies are used (Helietuha, H. H. Geleznaia, T. A., Kucheruk P. P. Oleinik, E. N., Triboi A. V., 2015). For example, in ultrasonic reactors, the intensification of heat and mass transfer of the esterification process reduces the quantitative composition of auxiliary substances and reduces the biodiesel yield time, but significantly increases the cost of production. Rotary-pulsation devices allow you to locally introduce energy and discretely distribute it relative to the impulse effect on the mixing of the components of a chemical reaction. This method reduces energy consumption for biodiesel production and reduces the duration of the process, but can cause pulsation of the working fluid flows, which complicates the esterification process and equipment efficiency.

The proposed systems to support the operation of biodiesel plants, including information, do not take into account the significant heat storage capacity of the oil and are based on measuring the oil temperature to change the oil flow to support the esterification process, which can upset the required flows balance of biodiesel and oil. For timely decision-making it is necessary to include in the biodiesel plant a heat exchanger for heating oil with biodiesel heat. It is necessary to predict the change in oil temperature by measuring the temperature of biodiesel at the outlet of the heat exchanger, which changes earlier than the change in oil temperature, and maintain the temperature of biodiesel at the inlet to the heat exchanger based on changes of heat exchange surface. The author, Chaikovskaya, E. (2016) proposes an integrated system for estimating changes in oil temperature obtained on the basis of mathematical and logical modeling as part of a cogeneration system is proposed, which allows to decide on changing the number of heat exchanger plates to ensure constant biodiesel output and timely change of heated and fresh oil.

Typically, biodiesel production requires up to 20% of the energy produced to support the esterification process for oil heating. By including a biodiesel plant in the cogeneration system using oil heating with biodiesel heat, it is necessary to maintain the local water temperature of the cooling circuit of the cogeneration system engine to ensure the ratio of electricity production and heat when changing consumption. t is proposed to predict the change in temperature of local water in the engine cooling circuit to maintain the ratio of electricity and heat production based on the change in the heat exchange surface of the engine cooling circuit during oil heating, shipment of heated oil and loading of fresh raw materials. The round-the-clock operation of biodiesel plants allows to extend the term of operation of cogeneration systems in terms of additional energy production with a guaranteed supply of biodiesel.

Methodological and Mathematical Substantiation

Thus, the cogeneration system on biodiesel – open integrated dynamic system, the operation of which requires maintaining the ratio of production and consumption of electricity and heat when when using the heat of biodiesel to heat the oil. The temperature of the oil depends on the ambient temperature, and the temperature of biodiesel affects the maintenance of oil heating.

The operation of the cogeneration system on biodiesel can be considered, in this regard, as the reproduction of external and internal influences and changes in initial conditions, such as changes in cogeneration power, changes in biodiesel temperature at the outlet of the heat exchanger, changes in cooling water temperature at the outlet of the heat exchanger of engine cooling, return water temperature, etc. The nature of the reaction is determined by the inertia of the devices and the rate of transients, i.e., the dynamic properties. Thus, the significant heat storage capacity of the heated oil increases the inertia of the transients, complicating the assessment of changes in oil temperature during measurement, so it must be predicted by evaluating analytically.

Dynamic properties are manifested in the operation of the cogeneration system, information about which can be obtained from the dynamic characteristics, such as changes in oil temperature at the outlet of the oil heating heat exchanger, local water temperature at the outlet of the engine cooling circuit. The dynamic characteristics of the cogeneration system with sufficient accuracy for practice can be described by a finite set of parameters for changes in time, and the spatial coordinate that coincides with the direction of flow of the medium, such as changes in oil temperature, local water at the outlet of the engine cooling circuit, etc. Input effects can also be described by a set of parameters, such as changes in the temperature of biodiesel at the outlet of the oil heating heat exchanger, changes in the temperature of the cooling water at the outlet of the engine cooling circuit, changes in return water temperature. Therefore, the dynamic description of the cogeneration system on biodiesel fuel most fully and multifacetedly characterizes its operation.

Thus, it is possible to determine that the real cogeneration system is a dynamic system, the mathematical model of which reflects the properties of the transformation of influences, i.e., its dynamic properties. Due to the fact that the cogeneration system reflects the dynamic peculiarities due to the nature of reactions to influences, the supporting of the functioning cogeneration system should be part of such a technological system, which is based on a dynamic system. Based on the methodological, mathematical, logical substantiation of the technological systems (Chapter 1), energy-saving technology of biodiesel production (the author, Chaikovskaya, E. (2016) the architecture, mathematical substantiation of the architecture (1), mathematical substantiation of maintenance of the operation (2) of the biodiesel cogeneration system Smart Grid are proposed (Figure 3.1).

A cogeneration system is a dynamic system, the operation of which is the reproduction of a change in external, internal influences and initial conditions, for example, changes in oil quality, biodiesel temperature, return water temperature, changes in power, changes in the ratio of electricity and heat onder at of changes in consumption etc. That is why, when designing a cogeneration biodiesel system, underlying which is an integrated dynamic subsystem (Figure 3.1). The integrated dynamic subsystem includes the following components: the electric network, the cogeneration unit, the engine cooling circuit as a part of the cogeneration plant, the biodiesel plant, the heat exchanger for heating oil with biodiesel heat. Other units included in the cogeneration system are the units of charge, discharge, and estimation of functional efficiency, which are in harmonized interaction with the integrated dynamic subsystem (Figure 3.1).

The mathematical substantiation of the architecture of the biodiesel cogeneration system Smart (1), (Figure 3.1), based on the methodology of the mathematical description of dynamics of power systems, the method of the graph of cause-effect relations [Chapter 1] is proposed (Figure 3.1).

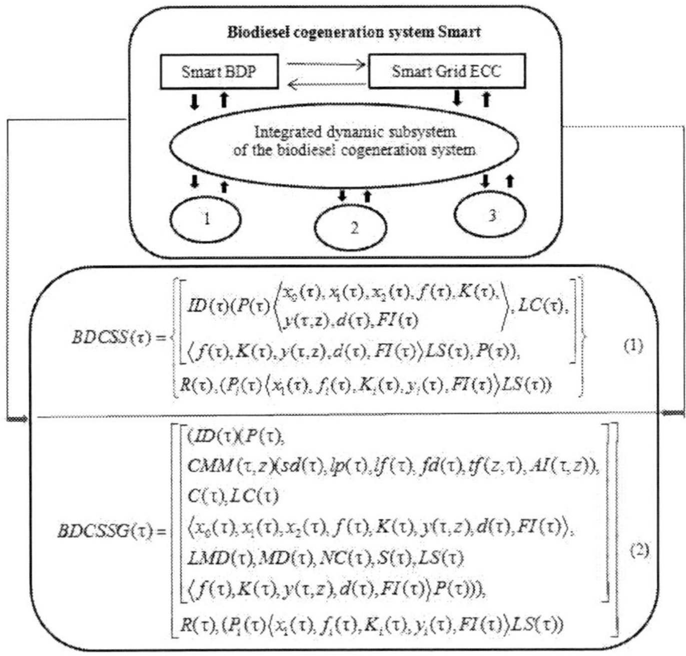

Figure 3.1. Biodiesel cogeneration system Smart Grid: the architecture: BDP – biodiesel plant; ECC – engine cooling circuit; 1– charge unit; 2–discharge unit; 3 – unit of evaluation of functional efficiency. Mathematical substantiation of the architecture (1). Mathematical substantiation of maintenance of the operation (2).

Where *BDCSS* (τ) – biodiesel cogeneration system; Smart; τ – time, s; *ID*(τ) – integrated dynamic subsystem (the electric network, the cogeneration unit; the engine cooling circuit as a part of the cogeneration plant, the biodiesel plant; the heat exchanger for heating oil with biodiesel heat); *P*(τ) – properties of the components of the cogeneration system; *x*(τ) – impacts (change in temperature of biodiesel at the outlet from the heat exchanger, temperature of cooling water at the outlet from the heat exchanger, temperature of return water)); *f*(τ) – parameters that are measured: (temperature of biodiesel at the outlet from the heat exchanger, temperature of cooling water at the outlet from the heat exchanger, temperature of return water; *K*(τ) – coefficients of mathematical description of the dynamics of temperature of oil, temperature of local water; *y*(τ, *z*) – output parameters (temperature of oil, temperature of local water); *z* – spatial coordinate of the oil heat exchanger, local water heat exchanger coincides with the direction of the flow of motion of the medium, m; *d*(τ) – dynamic parameters of temperature of oil, temperature of local water; $FI(\tau)$ – functional resulting information on decision making; $LC(\tau)$ – logical relations regarding the control of the cogeneration system workability; $LS(\tau)$ – logical relations regarding the identification of the state of the cogeneration system; $R(\tau)$ – logical relations in $BDCSS$ (τ) to confirm the correctness of decisions made from the units of the cogeneration system. Indices: i – the number of elements of the cogeneration system; 0, 1, 2 – initial stationary mode, external, internal nature of impacts.

The mathematical substantiation of maintenance of the operation of the biodiesel cogeneration system Smart Grid (2), (Figure 3.1), based on the methodology of the mathematical description of dynamics of power systems, the method of the graph of cause-effect relations (Chapter 1) is proposed. The basis of the proposed rationale is the mathematical description of the architecture of the biodiesel cogeneration system Smart (1), (Figure 3. 1). Prediction of a change in the temperature of oil and a change in the temperature of local water of the engine cooling circuit makes it possible to make forestalling decisions on a change of the number of the oil heat exchanger plates and of the engine cooling circuit to maintain the ratio of electricity production and heat when changing consumption. The temperature of the biodiesel at the outlet of the heat exchanger, the temperature of the cooling water at the outlet of the heat exchanger, and the temperature of return water are measured. Mathematical substantiation of Smart Grid maintenance of the operation of the biodiesel cogeneration system (2) is proposed (Figure 3. 1).

Where $BDCSSG$ (τ) – Smart Grid maintenance of the operation of the biodiesel cogeneration system; τ – time, s; $ID(\tau)$ – integrated dynamic subsystem (the electric network, the cogeneration unit, the engine cooling circuit as a part of the cogeneration plant, the biodiesel plant, the heat exchanger for heating oil with biodiesel heat. $P(\tau)$ – the properties of the elements of the integrated dynamic subsystem, units of the cogeneration system; $CMM(\tau, z)$ – complex mathematical modeling of the dynamics of changes in the temperature of oil, the temperature of local water; $sd(\tau)$ – the input data (productivity of the biodiesel plant and the type of the cogeneration system and its power; the temperature of oil at the inlet of the heat exchanger and at the outlet of the heat exchanger, $lp(\tau)$ – the boundary change in parameters (the temperature of biodiesel at the inlet of the heat exchanger and at the outlet of the heat exchanger, the temperature of oil at the inlet of the heat exchanger and at the outlet of the heat exchanger, the temperature of cooling water at the inlet of the heat exchanger and at the outlet of the heat exchanger, the temperature of local water at the inlet of the heat exchanger and at the outlet of the heat exchanger, the temperature of return water; *lf*(τ) the levels of operation of the biodiesel plant, cogeneration system; *fd*(τ) – the obtained parameters (parameters of heat exchange in the heat exchanger of the oil, engine cooling circuit, time constants and coefficients of the mathematical model of dynamics of a change in the temperature of oil, the temperature of local water; *tf*(τ,*z*) – the transfer function of predicted parameters – the temperature of oil, the temperature of local water; *AI*(τ,*z*) – the standard information regarding the evaluation of the maximum admissible change in the temperature of oil, the temperature of local water;*C*(τ) – the control of workability of the cogeneration system; *LC*(τ) – the logical relations of the control of the cogeneration system workability; *x*(τ) – impacts (change in temperature of biodiesel at the outlet from the heat exchanger, temperature of cooling water at the outlet from the heat exchanger, temperature of return water); *f*(τ) – the measured parameters: (temperature of biodiesel at the outlet from the heat exchanger, temperature of cooling water at the outlet from the heat exchanger, temperature of return water); *K*(τ) – the coefficients of the mathematical description of the dynamics of a change in the temperature of oil, the temperature of local water; *y*(τ, *z*) – the output parameters (the temperature of oil, the temperature of local water); *z* – the coordinate of the length of the heat exchanger of the oil, the heat exchanger of the cooling circuit, m; *d*(τ) – the dynamic parameters of estimation of a change in the temperature of oil, the temperature of local water; *FI*(τ) – functional resulting information on decision making; *LMD*(τ) – the logical relations of decision making; *MD*(τ) – decision making; *NC*(τ) – the new conditions of the cogeneration system operation; *S*(τ) – the identification of the state of the cogeneration system; *LS*(τ) – the logical relations of identification of the state of the cogeneration system; *R*(τ) – the logical relations between the dynamic subsystem and units of charge, discharge, functional estimation of efficiency that belong to the cogeneration system. Indices: *i* – the number of elements of *BDCSSG* (τ); 0, 1, 2 – the initial, external, and internal character of influences.

Mathematical substantiation of the architecture of the biodiesel cogeneration system Smart (1) and mathematical substantiation of maintenance of the operation of the biodiesel cogeneration system Smart Grid (2) (Figure 3.1) make it possible to maintain the operation of the cogeneration system using the following actions:

- Workability control (*C*(τ)) of the dynamic subsystem based on complex mathematical (*CMM*(τ, *z*)) and logical (*LC*(τ)) modeling regarding obtaining standard ($AI(\tau,z)$) estimate of a change in the temperature of oil of a change in the temperature of local water;

- Workability control ($C(\tau)$) of the dynamic system based on complex mathematical ($CMM(\tau, z)$) and logical ($LC(\tau)$) modeling regarding the obtaining functional ($FI\ (\tau)$) estimate of a change in the temperature of oil, of a change in the temperature of local water;
- Decision making ($MD(\tau)$) with the use of the functional resulting information ($FI\ (\tau)$), obtained based on logical modeling ($LMD(\tau)$); decision making to maintain the ratio of production and consumption of electric power and heat with the use of the functional assessment of a change in the temperature of oil, of a change in the temperature of local water;
- Identification ($S(\tau)$) of the new conditions of functioning of the cogeneration system ($NC(\tau)$) based on logical modeling ($LS(\tau)$) as a part of the dynamic subsystem and confirmation of new operating conditions based on logical modeling ($R(\tau)$) from the units of the cogeneration system.

Complex Mathematical and Logical Modeling of the Biodiesel Cogeneration System Smart Grid

According to Formulas (1), (2), the prediction of a change in the temperature of oil is proposed. The temperature of biodiesel at the outlet from the heat exchanger is measured. The transfer function along the channel "temperature of oil – temperature of biodiesel" is obtained. A change in the oil temperature is estimated both over time and along the spatial coordinate of the heat exchanger coincides with the direction of the flow of motion of the medium.

The transfer function along the channel " temperature of oil – temperature of biodiesel," which was obtained as a result of solving a system of nonlinear differential equations, is presented as follows:

$$W_{t-\vartheta_1} = \frac{K_{\text{bd}}\varepsilon\left(1 - L^*_{\text{bd}}\right)}{\left(T_{\text{oil}}S + 1\right)\beta \text{ - } 1}\left(1 - e^{-\gamma\xi}\right), \tag{3}$$

where

$$K_{bd} = \frac{m(\theta_0 - \sigma_0)}{G_{bd0}}; \; \varepsilon = \frac{\alpha_{bd0} h_{bd0}}{\alpha_{oil0} h_{oil0}}; \; L_{bd} = \frac{G_{bd} C_{bd}}{\alpha_{bd0} h_{bd0}};$$

$$T_{oil} = \frac{g_{oil} C_{oil}}{\alpha_{oil0} h_{oil0}}; \; \beta = T_m S + \varepsilon^* + 1; \; T_m = \frac{g_m C_m}{\alpha_{oil_0} h_{oil0}}; \; \varepsilon^* = \varepsilon(1 - L_{bd}^*);$$

$$\gamma = \frac{(T_{oil} S + 1)\beta - 1}{\beta}; \; \xi = \frac{z}{L_{oil}}; \; L_{oil} = \frac{G_{oil} C_{oil}}{\alpha_{oil0} h_{oil0}},$$

where C is the specific thermal capacity, kJ/(kg·K); α is the heat transfer factor, kW/(m^2·K); G is the consumption of substance, kg/s; g is the specific weight of a substance, kg/m; h is the specific surface, m^2/m; t, σ, θ are the temperature of the oil, biodiesel and of the separating wall, respectively, K; z is spatial coordinate of the heat exchanger coincides with the direction of the flow of motion of the medium, m; $T_{i,}$, T_m are the time constants that characterize the thermal accumulating capacity of oil, metal, s; m is the indicator of the dependence of heat transfer factor on consumption; τ is the time, s; S is the Laplace parameter; $S = \omega j$; ω is the frequency, 1/s.; Indices: 0 – initial stationary mode; 1 – inlet to the heat exchanger; oil – internal flow – oil; bd – external flow – biodiesel; m – metal wall.

Transfer function along the channel: "temperature of oil – temperature of biodiesel" is obtained based on the solution of a system of nonlinear differential equations using the Laplace transform tool. The system of differential equations includes the equation of state as the estimation of the physical model of the biodiesel plant. The system of differential equations also includes the equation of energy of transmitting and receiving media – biodiesel and oil, respectively, and the equation of thermal balance for the wall of the oil heating heat exchanger. The equation of the energy of the receiving medium is developed with the representation of a change in oil both in time and along the spatial coordinate, which coincides with the direction of the motion flow of the medium. The equation of the energy of the transmitting medium includes the K_{bd} coefficient, which assesses a change in the temperature of biodiesel at the outlet of the heat exchanger that is measured.

A real part of the transfer function is separated:

$$O(\omega) = \frac{(L_1 A_1) + (M_1 B_1)(1 - L_3^*)}{(A_1^2 + B_1^2)}. \tag{4}$$

The K_{bd} factor includes the temperature of the separating wall θ:

$$\theta = (\alpha_{oil}(\sigma_1 + \sigma_2)/2) + (A(t_1 + t_2)/2)/(\alpha_{oil} + A), \tag{5}$$

where σ_1, σ_2 are the temperatures of biodiesel at the inlet and at the outlet of the heat exchanger, K, respectively; t_1, t_2 are the temperatures of oil at the inlet and at the outlet of the heat exchanger, K, respectively; α is the heat transfer factor, kW/(m^2·K). Indice oil – internal flow: oil.

$$A = 1/(\delta_m / \lambda_m + 1/\alpha_{bd}), \tag{6}$$

where δ is the thickness of a wall of the heat exchanger, m; λ is the thermal conductivity of the metal wall of the heat exchanger, kW/(m·K). Indices: bd – external flow: biodiesel; m – metal wall of a heat exchanger.

To use the real part $O(\omega)$, the following factors were obtained:

$$A_1 = \varepsilon^* - T_{oil} T_m \omega^2; \tag{7}$$

$$A_2 = \varepsilon^* + 1; \tag{8}$$

$$B_1 = T_{oil}\varepsilon^*\omega + T_{oil}\omega + T_m\omega; \tag{9}$$

$$B_2 = T_m\omega; \tag{10}$$

$$C_1 = \frac{A_1 A_2 + B_1 B_2}{A_2^2 + B_2^2}; \tag{11}$$

$$D_1 = \frac{A_2 B_1 - A_1 B_2}{A_2^2 + B_2^2}; \tag{12}$$

$$L_1 = 1 - e^{-\xi C_1}\cos(-\xi D_1); \tag{13}$$

$$M_1 = -e^{-\xi C_1}\sin(-\xi D_1). \tag{14}$$

The transfer function (3), which was obtained based on the use of the operator method of solving the system of nonlinear differential equations, retains the Laplace transform parameter – S ($S = \omega j$), where ω is the frequency, 1/s. To switch from the frequency area to the time area, a real part (4), obtained as a result of the mathematical treatment of transfer function, was separated. It is this part that is included in the integrals (15), which makes it possible to obtain dynamic characteristics of a change the temperature of oil using the inverse Fourier transform:

$$t(\tau, z) = \frac{1}{2\pi}\int_0^\infty K_{bd} O(\omega)\sin(\tau\omega/\omega)d\omega, \tag{15}$$

where t is the temperature of oil, K.

According to formulas (1), (2), the prediction of a change in the temperature of local water is proposed. In the engine cooling circuit, the following parameters are measured: the temperature of the cooling water at the outlet of the heat exchanger and the return water temperature. The transfer function along the channel "temperature of local water – temperature of cooling water " is obtained. A change in the local water temperature is estimated both over time and along the spatial coordinate of the heat exchanger coincides with the direction of the flow of motion of the medium. The transfer function along the channel " temperature of local water – temperature of cooling water," which was obtained as a result of solving a system of nonlinear differential equations, is presented as follows:

$$W_{t-\vartheta_1} = \frac{K_{cw}\varepsilon(1 - L_3^*)}{(T_{lw}S + 1)\beta - 1}(1 - e^{-\gamma\xi}), \tag{16}$$

where

$$K_{cw} = \frac{m(\theta_0 - \sigma_0)}{G_{cw0}}; \quad \varepsilon = \frac{\alpha_{cw0}h_{cw0}}{\alpha_{lw0}h_{lw0}}; \quad L_{cw} = \frac{G_{cw}C_{cw}}{\alpha_{cw0}h_{cw0}};$$

$$T_{\mathrm{lw}} = \frac{g_{\mathrm{lw}} C_{\mathrm{lw}}}{\alpha_{\mathrm{lw0}} h_{\mathrm{lw0}}};\quad \beta = T_{\mathrm{m}} S + \varepsilon^{*} + 1;\quad T_{\mathrm{m}} = \frac{g_{\mathrm{m}} C_{\mathrm{m}}}{\alpha_{\mathrm{lw0}} h_{\mathrm{lw0}}};\quad \varepsilon^{*} = \varepsilon\left(1 - L_{cw}^{*}\right);$$

$$\gamma = \frac{(T_{\mathrm{lw}} S + 1)\beta - 1}{\beta};\quad \xi = \frac{z}{L_{\mathrm{lw}}};\quad L_{\mathrm{lw}} = \frac{G_{\mathrm{lw}} C_{\mathrm{lw}}}{\alpha_{\mathrm{lw0}} h_{\mathrm{lw0}}},$$

where C is the specific thermal capacity, kJ/(kg·K); α is the heat transfer factor, kW/(m^2·K); G is the loss of substance, kg/s; g is the specific weight of a substance, kg/m; h is the specific surface, m^2/m; t, σ, θ are the temperature of the local water, cooler water and of the separating wall, respectively, K; z is spatial coordinate of the heat exchanger coincides with the direction of the flow of motion of the medium, m; T_{lw}, T_{m} are the time constants that characterize the thermal accumulating capacity of local water, metal, s; m is the indicator of the dependence of heat transfer factor on consumption; τ is the time, s; S is the Laplace parameter; $S = \omega j$; ω is the frequency, 1/s.; Indices: 0 – initial stationary mode; 1 – inlet to the heat exchanger; lw – internal flow – local water; cw – external flow – cooler water; m – metal wall.

Transfer function along the channel: “local water temperature - cooling water temperature” was obtained based on the solution of a system of nonlinear differential equations using the Laplace transform tool. The system of differential equations includes the equation of state as the estimation of the physical model of the cogeneration system. The system of differential equations also includes the equation of energy of transmitting and receiving media – cooling water and local water, respectively, and the equation of thermal balance for the wall of the heat exchanger of the engine cooling circuit. The equation of the energy of the receiving medium is developed with the representation of a change in local water both in time and along the spatial coordinate, which coincides with the direction of the motion flow of the medium. The equation of the energy of the transmitting medium includes the coefficient K_{cw}. which assesses a change in the temperature of cooling water at the outlet of the heat exchanger and the temperature of return water are measured.

A real part of the transfer function is separated:

$$O(\omega)=\frac{(L_1A_1)+(M_1B_1)(1-L_3^*)}{(A_1^2+B_1^2)}. \tag{17}$$

The K_{cw} factor includes the temperature of the separating wall θ:

$$\theta=(\alpha_{lw}(\sigma_1+\sigma_2)/2)+(A(t_1+t_2)/2)/(\alpha_{lw}+A), \tag{18}$$

where σ_1, σ_2 are the temperatures of cooling water at the inlet and at the outlet of the heat exchanger, K, respectively; t_1, t_2 are the temperatures of local water at the inlet and at the outlet of the heat exchanger, K, respectively; α is the heat transfer factor, kW/(m^2·K). Index lw – internal flow: local water

$$A=1/(\delta_m/\lambda_m+1/\alpha_{cw}), \tag{19}$$

where δ is the thickness of a wall of the heat exchanger, m; λ is the thermal conductivity of the metal wall of the heat exchanger, kW/(m·K). Indices: cw – external flow: cooling water; m – metal wall of a heat exchanger.

To use the real part $O(\omega)$, the following factors were obtained:

$$A_1=\varepsilon*-T_{lw}T_m\omega^2; \tag{20}$$

$$A_2=\varepsilon^*+1; \tag{21}$$

$$B_1=T_{lw}\varepsilon^*\omega+T_{lw}\omega+T_m\omega; \tag{22}$$

$$B_2=T_m\omega; \tag{23}$$

$$C_1=\frac{A_1A_2+B_1B_2}{A_2^{\,2}+B_2^{\,2}}; \tag{24}$$

$$D_1=\frac{A_2B_1-A_1B_2}{A_2^{\,2}+B_2^{\,2}}; \tag{25}$$

$$L_1 = 1 - e^{-\zeta C_1} \cos\left(-\xi D_1\right); \tag{26}$$

$$M_1 = -e^{-\zeta C_1} \sin\left(-\xi D_1\right). \tag{27}$$

The transfer function (16), which was obtained based on the use of the operator method of solving the system of nonlinear differential equations, retains the Laplace transform parameter – S ($S = \omega j$), where ω is the frequency, 1/s. To switch from the frequency area to the time area, a real part (17), obtained as a result of the mathematical treatment of transfer function, was separated. It is this part that is included in the integrals (28), which makes it possible to obtain dynamic characteristics of a change the temperature of local water using the inverse Fourier transform:

$$t\left(\tau, z\right) = \frac{1}{2\pi} \int_0^\infty K_{cw} O\left(\omega\right) \sin\left(\tau\omega/\omega\right) d\omega, \tag{28}$$

where t is the temperature of local water, K.

Thus, for obtaining the reference estimation of change oil temperature and temperature of local water a block diagram (Figure. 3.2) is proposed using, for example, the initial data of the biodiesel cogeneration system based on engine type *MAN D*2840 *LE*201 total power 451 kW, electrical power 271.5 kW, heat power 175.9 kW.

The biodiesel cogeneration system based on engine type *MAN D*2840 *LE*201 total power 451 kW, electrical power 271.5 kW, heat power 175.9 kW includes a biodiesel unit type EXON - 1000, with a capacity of 24000 l/day of biodiesel. Constant consumption of biodiesel– 0.245 kg/s and constant consumption of oil – 0.254 kg/s. The following levels of operation of the biodiesel plant have been established for the change in the temperature of the biodiesel at the inlet to the heat exchanger and at the outlet from the heat exchanger: – first level: 54–34.16°C; second level: 50–30°C; third level: 46–25.5°C. They correspond to changes in the number of plates of the heat exchanger: 22, 18, 14 to support the heating of the oil from 20°C to 45°C. The following levels of operation of the cogeneration system have been established for the change in the temperature of the cooling water at the inlet to the heat exchanger and at the outlet from the heat exchanger: first level: 95–75°C; second level: 95–92.78°C; third level: 95–70°C; fourth level: 95 – 66.43°C for heating 2.59 kg/s of local water from 30°C to 55°C. They correspond to changes in the number of plates of the heat exchanger: 46, 40, 34, 28 and to changes in cooling water consumption: 3.24 kg/s, 2.92 kg/s, 2.59 kg/s, 2.27 kg/s.

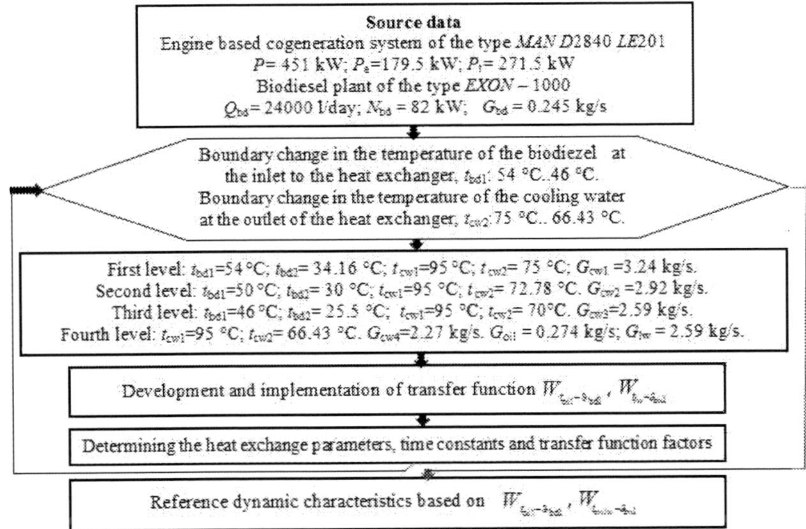

Figure 3.2. Block diagram of complex mathematical modeling of the cogeneration biodiesel system Smart Grid: P, P_e, P_t are the nominal, electric, thermal power of the cogeneration system, respectively, kW; Q_{bd} – productivity of biodiesel plant, l/day; N – electric power of the biodiesel plant, kW; G_{bd}, G_{oil}, G_{cw}, G_{lw} are the consumption biodiesel, oil, cooling water, local water, respectively, kg/s; t_{bd1}, t_{bd2}, t_{cw1}, t_{cw2} are the temperature of biodiesel at the inlet of the heat exchanger and at the outlet of the heat exchanger, the temperature of cooling water at the inlet of the heat exchanger of the engine cooling circuit and at the outlet of the heat exchanger, respectively, °C.

According to formulas (1) - (3), (16) and the proposed block diagram (Figure 2.3), the results of reference information obtained on the basis of complex mathematical modeling of the biodiesel cogeneration system Smart Grid is presented (Tables 3.1–3.4).

Table 3.1. Heat transfer parameters as part of complex mathematical modeling of oil heat exchanger

Levels of operation	Parameter		
	α_{bd}, kW/(m²·K)	α_{oil}, kW/(m²·K)	k, kW/(m²·K)
First level	0.586	0.302	0.198

Second level	0.664	0.345	0.224
Third level	0.930	0.409	0.281

Note: α_{bd} – coefficient of convective heat transfer from the biodiesel to the heat exchanger wall, kW/(m²·K); α_{oil} – coefficient of convective heat transfer from the heat exchanger wall to oil, kW/(m²·K); *k* – heat transfer coefficient, kW/(m²·K).

Table 3.2. Time constants and coefficients of the mathematical model of the dynamics of the temperature of the oil

Levels of operation	T_{oil}, c	T_m, c	ε	ε*	L_{bd},M	L_{oil}, M	L_{bd}^*	ζ
First level	9.28	6.81	2.35	2.32	86.0	159.96	0.011	1. 13
Second level	8, 13	5.96	2.33	2,30	75.28	140.05	0.013	1.29
Third level	6, 85	5.03	2.31	2.27	52.38	118.13	0.019	1.41

Table 3.3. Heat transfer parameters as part of complex mathematical modeling of local water heat exchanger

Levels of operation	Parameter		
	α_{lw}, kW/(m²·K)	α_{cw}, kW/(m²·K)	k, kW/(m²·K)
First level	11.105	8.303	4.137
Second level	11.411	9.118	4.375
Third level	11.786	10.166	4.663
Fourth level	12.240	11.579	5.018

Note: α_{cw} – coefficient of convective heat transfer from the cooling water to the heat exchanger wall, kW/(m²·K); α_{lw} – coefficient of convective heat transfer from the heat exchanger wall to local water, kW/(m²·K); *k* – heat transfer coefficient, kW/(m²·K).

Table 3.4. Time constants and coefficients of the mathematical model of the dynamics of the temperature of the local water

Levels of operation	$T_{lw, s}$	$T_{m, s}$	ε	ε*	$L_{cw, m}$	$L_{lw, m}$	$L_{cw,}^*$	ζ
First level	0.595	0.217	1. 62	1.60	68.33	88.62	0.0144	0.649
Second level	0,542	0,198	1.52	1.50	59.85	80.69	0.0164	0.680
Third level	0,486	0,177	1.41	1.38	51.51	72.37	0.0190	0.720
Fourth level	0,427	0,156	1.28	1.25	43.40	63.54	0.0225	0.772

Presented in Tables 3.2, 3.4 time constants and coefficients that are part of the mathematical models of dynamics (3),(16) are obtained on the basis of parameters as part of complex mathematical modeling (Tables 3.1, 3.3).

Based on the proposed mathematical substantiation Smart Grid maintenance of functioning of the cogeneration system (1) to (3), (16) the block diagram for obtaining functional estimation of a change in the temperature of oil and the temperature of local water based the control of workability of the cogeneration on biodiesel fuel (Figure 3.3) is developed.

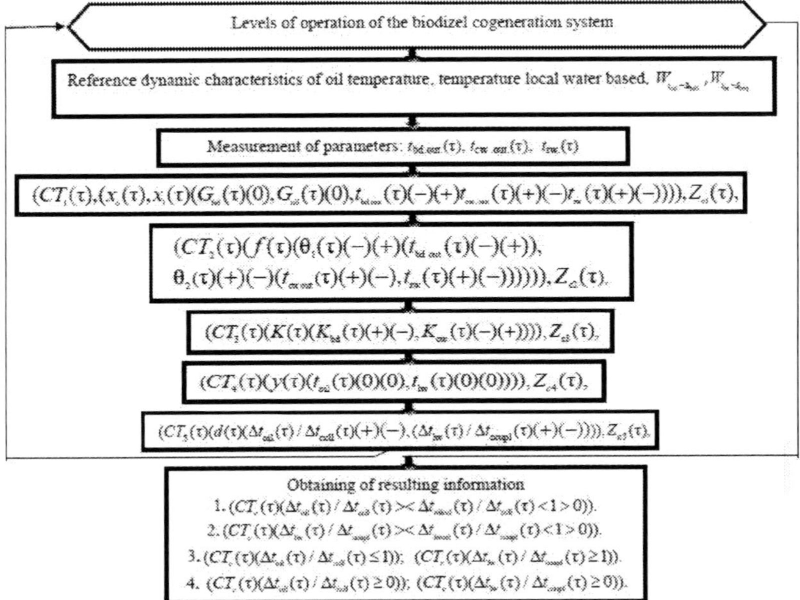

Figure 3.3. Block diagram of control of workability of the cogeneration system: $t_{bd.out.}(\tau)$, $t_{cw.out.}(\tau)$, $t_{rw.}(\tau)$ are the temperatures of biodiesel at the outlet of the heat exchanger, cooling water at the outlet of the heat exchanger of the engine cooling circuit, the temperature of return water, respectively, °C; θ_1, θ_2 are the temperatures of the separating wall of the oil heat exchanger, of the heat exchanger of the engine cooling circuit, respectively, °C; *CT* is the event control; *Z* is the logical relations; *d* is the dynamic parameters; *x* is the influences; *f* is the diagnosed parameters; *y* is the output parameters; *K* is the mathematic description factors; ι is time. Indices: *c* is efficiency control; bd is biodiesel; 1,2 constant, calculated value of the parameter of the low, upper level of functioning 0, 1, 2 are the initial stationary mode, external, internal parameters; 3 is the coefficients of dynamics equations; 4 is the essential diagnosed parameters; 5 is dynamic parameters.

Control of workability of the cogeneration system (Figure 3.3) enables obtaining the resulting information on decision-making about the maintenance of the functioning of the cogeneration system.

Smart Grid System of Maintaining the Operation of the Cogeneration Biodiesel System at the Decision-Making Level

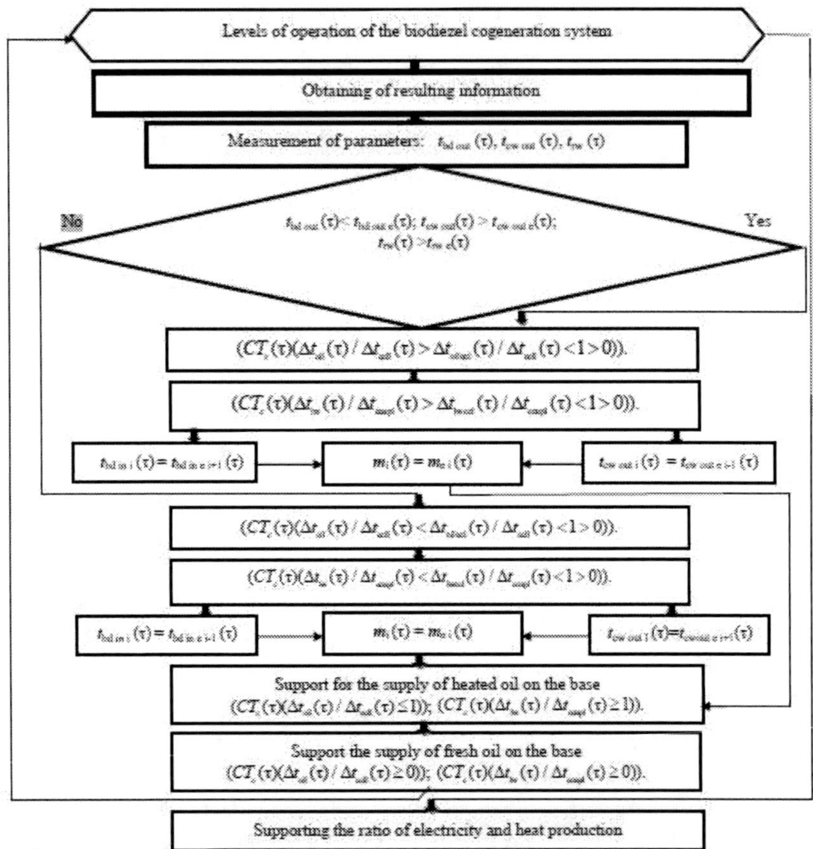

Figure 3.4. Block diagram of maintenance of functioning of the cogeneration system: CT_c is the efficiency control; t_{oil}, t_{bd}, t_{cw}, t_{rw} are the temperature of oil, biodiesel, cooling water, local water, return water, respectively, K; ι is the time; m is the ratio of electricity and heat production. Indices: *i* is the number of operation levels; e is the reference value of the parameter; in, outlet are the inlet of the heat exchanger, the outlet of the heat exchanger, respectively; ccupl, ccll, are the constant calculation value of the parameter of the upper, low level of functioning respectively; ccl is the constant calculation value of the parameter of the level of functioning temperature cooling water at the inlet of the heat exchanger of the engine cooling circuit, at the outlet of the heat exchanger, the of return water.

Table 3.5. Integrated Smart Grid System of harmonization of production and consumption of electric power and heat

Time, τ, 10^2 s	Changing the parameters of the technological process	$\Delta t_{oil}(\tau)/\Delta t_{1}(\tau)$	$t_{oil}(\tau)$, °C	$\Delta t_{lw}(\tau)/\Delta t_2(\tau)$	$t_{lw}(\tau)$, °C
11	Serving fresh oil for heating t_1= 54°C; t_2= 42°C; t_3 =95°C; t_4 = 66.43°C; t_{rw} =30°C; n_1= 22 pieces; n_2= 28 pieces; G=2.27 kg/s; Q_{T1}= 10.7 kW; Q_{T2}=271.7 kW; P_T=261 kW; P_e=172.4 kW; m=0.6606	0.058	24.17	0.086	32.15
22	Charge t_1= 54°C; t_2= 37°C; t_3 =95° C; t_4 = 67°C; $t_{rw.}$= 30.5°C; n_1= 22 pieces; n_2= 28 pieces; G=2.27 kg/s; Q_{T1}= 15.1 kW; Q_{T2}=266.3 kW; P_T=251.2 kW; P_e=169.3 kW; m=0.6740	0.3057	27.64	0.433	40.825
33	Charge t_1= 54°C; t_2= 36°C; t_3 =95°C; t_4 =68°C; $t_{rw.}$= 30.75°C; n_1= 22 pieces; n_2= 28 pieces; G=2.27 kg/s; Q_{T1}=16 kW; Q_{T2}=256.8 kW; P_T=240.8 kW; P_e=168.3 kW; m=0.6990	0.3558	28.89	0.6	45
44	Decision making n_1= 18 pieces; n_2= 34 pieces; G=2.59 kg/s; t_1= 54°C; t_2= 34°C; t_3 =95°C; t_4 =70°C; $t_{rw.}$= 31°C; Q_{T1}=17.8 kW; Q_{T2}=271.3 kW; P_T=253.5; kW P_e= 167.7 kW; m=0.6617	0.4557	31.39	0.68	47
55	Identification of the new operating conditions n_1= 18 pieces; n_2= 34 pieces; G=2.59 kg/s; t_1= 50°C; t_2 = 33°C; t_3 =95°C; t_4 =70°C;	0.4557	31.39	0.68	47

Table 3.5. (Continued)

Time, τ, 10^2 s	Changing the parameters of the technological process	$\Delta t_{oil}(\tau)/\Delta t_1(\tau)$	$t_{oil}(\tau)$, °C	$\Delta t_{lw}(\tau)/\Delta t_2(\tau)$	$t_{lw}(\tau)$, °C
	$t_{rw.}$= 31°C; Q_{T1}=15.1 kW; Q_{T2}=271.3 kW; P_T=256.2 kW; P_e= 169.5 kW; m=0.6616				
66	Charge t_1= 50°C; t_2= 32°C; t_3 =95°C; t_4 = 71°C; $t_{rw.}$= 31.5°C; n_1= 18 pieces; n_2= 34 pieces; G=2.59 kg/s; Q_{T1}=16 kW; Q_{T2}=260.4 kW; P_T=244.4 kW; P_e =168.5 kW; m=0.6893	0.4967	32.46	0.822	50.55
77	Charge t_1= 50°C; t_2= 30°C; n_1= 18 pieces. Decision making n_2= 40 pieces; G=2.92 kg/s; t_3 =95° C; t_4= 72.78°C; $t_{rw.}$= 32°C; Q_{T1}=17.8 kW; Q_{T2}=271.9 kW; P_T=254.1 kW; P_e=167.7 kW; m=0.6600	0.6058	35.19	0.8458	51.145
88	Decision making n_1= 14 pieces; t_1= 50°C; t_2= 29°C; Identification of the new operating conditions t_3 =95°C; t_4= 72.78°C; $t_{rw.}$= 32°C; n_2= 40 pieces; G=2.92 kg/s; Q_{T1}=18.7 kW; Q_{T2}=271.9 kW; P_T=253.2 kW; P_e=167.1 kW; m=0.6602	0.6603	36.55	0.8458	51.145
99	Charge n_1= 14 pieces; t_1= 46°C; t_2= 27°C; n_2= *40* pieces; t_3 =95°C; t_4= 73.2°C; G=2.92 kg/s; $t_{rw.}$= 32°C; Q_{T1}=16,9 kW; Q_{T2}=266.7 kW; P_T=249.8 kW P_e=*168.1* kW; m=0.6730	0.8829	42.36	0.927	53.175

Time, τ, 10^2 s	Changing the parameters of the technological process	$\Delta t_{oil}(\tau)/\Delta t_{1}(\tau)$	$t_{oil}(\tau)$, °C	$\Delta t_{lw}(\tau)/\Delta t_2(\tau)$	$t_{lw}(\tau)$, °C
110	Supply of heated oil for esterification n_1= 14 pieces; t_1= 46°C; t_2= 25.5°C; Q_{T1}=18.2 kW; Decision making n_2= 46 pieces; t_3 =95°C; t_4= 75°C; G=3.24 kg/s; $t_{rw.}$= 32.25°C; Q_{T2}=271.5 kW; P_T=253.3 kW; P_e=167.5 kW; m=0.6612	1	45	0.93	53.25
121	Identification of the new operating conditions t_3 =95°C; t_4= 75°C; G=3.24 kg/s; $t_{rw.}$= 32.25°C; Q_{T1}=18.2 kW; Q_{T2}=271.5 kW; P_T=253.3 kW; P_e=167.5 kW; m=0.6612	1	45	0.93	53.25
Time, τ, 10^2 s	Changing the parameters of the technological process	$\Delta t_{oil}(\tau)/\Delta t_{1}(\tau)$	$t_{oil}(\tau)$, °C	$\Delta t_{lw}(\tau)/\Delta t_2(\tau)$	$t_{lw}(\tau)$, °C
132	Support for the supply of heated oil for etherification: t_3 =95°C; t_4= 75°C; G=3.24 kg/s; $t_{rw.}$= 30°C; Q_{T1}=17.8 kW; Q_{T2}=271.5 kW; P_T=253.7 kW; P_e=167.7 kW; m=0.6611	1	45	1	55

Note: t_{oil}, t_{lw}, t_{rw} are the oil temperature, local water temperature, return water temperature, °C; t_1, t_2 are biodiesel temperature at the inlet to the heat exchanger, at the outlet from the heat exchanger, respectively, °C; t_3, t_4 – cooling water temperature at the inlet to the heat exchanger, at the outlet from the heat exchanger, respectively, °C; Q_{T1}, Q_{T2} are the thermal power of the oil heat exchanger, cooling heat exchanger, respectively, kW; G is cooling water consumption, kg/s; n_1, n_2 are the number of plates of the oil heat exchanger, of the engine cooling circuit. respectively; P_T, P_e are the active thermal, electric power of the cogeneration system, respectively, kW; m is the ratio of production and consumption of electric power and heat; τ is time, s. Indexes: oil, lw, rw are oil, local water, return water; 1,2 constant, calculated value of the parameter of the low, upper level of functioning.

Based on the proposed mathematical substantiation (1) to (3), (16), the resulting information on decision-making (Figure 3.3) the block diagram of the maintenance of the functioning of the cogeneration biodiesel system (Figure 3.4) are presented.

Results and Discussion

The comprehensive integrated system of maintenance of operation of the cogeneration biodiesel system is developed. (Table 3.5). There is a continuous measurement of the temperature of biodiesel at the outlet from the heat exchanger of oil heating. In the engine cooling circuit the temperature of cooling water at the outlet of the heat exchanger and the temperature of return water are measured.

The temperature of oil in the established period is determined as follows:

$$t_{\text{oil}\,i+1}(\tau) = t_{\text{oil}\,i} + \\ +\begin{pmatrix} \Delta t_{\text{oil}\,i+1}(\tau) / \Delta t_{\text{oilccl.}}(\tau) - \\ -\Delta t_{\text{oil}\,i}(\tau) / \Delta t_{\text{oilccl.}}(\tau) \end{pmatrix} \left(t_{\text{oil2}} - t_{\text{oil1}}\right), \tag{29}$$

where t_{oil} is temperature of oil, °C; $t_{1\text{oil}}$, $t_{2\text{oil}}$ are the initial, final values of temperature of oil, °C; τ is the time, s. Index: ccl is the constant, calculated value of the parameter of the lower level of functioning; *i* is the number of levels of the biodiesel plant.

A temperature of the local water in the established period is determined as follows (табл. 3.5):

$$t_{lw\,i+1}(\tau) = t_{lw\,i} + \\ +\begin{pmatrix} \Delta t_{lw\,i+1}(\tau) / \Delta t_{\text{ccup.}}(\tau) - \\ -\Delta t_{lw\,i}(\tau) / \Delta t_{\text{ccup}}(\tau) \end{pmatrix} \left(t_{lw2} - t_{lw1}\right), \tag{30}$$

where t_{lw} is temperature of local water, °C; $t_{1\text{lw}}$, $t_{2\text{lw}}$ are the initial, final values of temperature of local water, °C; τ is the time, s. Index: ccup is the constant, calculated value of the parameter of the upper level of functioning; *i* is the number of levels of the cogeneration system.

So, for example, in the period of time $44 \cdot 10^2$ s after serving fresh oil for heating predicting the maintenance of heating oil to the level of 31.39°C in this time period (Table 3.5) the forestalling decision to decrease the number of plates of the heat exchanger of the oil heating from 22 pieces to 18 pieces was made.

The absolute value of the temperature of oil (Table 3.5) with the use of formula (29) in the period $44 \cdot 10^2$ s (1.8 hours) is:

$$31.39°C = ((0.4557\text{-}0.3558)(45°C - 20°C)) + 28{,}89°C.$$

Predicting the maintenance of heating local water to the level of 47 °C in this time period (Table 3.5), the forestalling decision to increase the number of plates of the heat exchanger of the engine cooling circuit from 28 pieces to 34 pieces was made.

The absolute value of the temperature of local water (Table 3.5) with the use of formula (30) in the period $44 \cdot 10^2$ s (1.8 hours) is:

$$47°C = 45°C + (0.68 - 0.6)(55 - 30°C).$$

The ratio of production of electric power and heat at a change in consumption is maintained at the level of 0.6653 (Table 3.5, Figure 3.5).

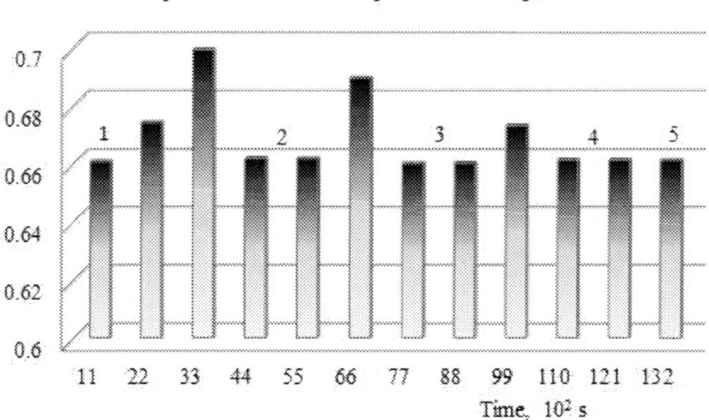

Figure 3.5. Support for the ratio of production of electric power and heat at a change of consumption , where 1 – support for the supply of serving fresh oil for heating; 2, 3, 4 – decision making for the change in the number of plates of the biodiesel heat exchanger and the heat exchanger of the engine cooling circuit; 5 – support for the supply of heated oil for etherification.

And if the temperature of biodiesel at the outlet of the heat exchanger decreased to 25.5°C, then due to the completion of heating the oil to 45°C, it is necessary to supply heated oil to the biodiesel reactor and load fresh oil, including 22 plates of heat exchanger and 28 circuit plates engine cooling to maintain the ratio of electricity production and heat when changing consumption at the level of 0.6877.

Chapter 4

Integrated Smart Grid System for Maintaining the Functioning of a Drying Plant as a Part of a Cogeneration System

Abstract

The integrated Smart Grid System of harmonization of production and consumption of electric power and heat was developed. The integrated dynamic subsystem includes the following components: the electric network, the cogeneration plant, the second circuit as a part of the cogeneration plant, the drying chamber, the heat exchanger for heating air with local water of second circuit of the cogeneration plant, air fan.

A change in the ratio of production and consumption of electric power and heat, change in the moisture content of the air in the drying chamber and change in the temperature of local water of the second circuit of the cogeneration system is predicted by measuring temperature of air at the outlet from the heat exchanger, the temperature of the of flue gases at the inlet of the heat exchanger and the return water temperature. Making forestalling decisions to support the supply of raw timber for drying, dried timber for production of the pellet fuel, changing number of the number of revolutions of the electric motor of the air fan and the changing number of plates of the heat exchanger of the second circuit makes it possible to maintain the ratio of production and consumption of electric power and heat and the moisture content of the air and local water

The complex mathematical and logical modeling of the cogeneration system, based on the mathematical substantiation of the architecture of the cogeneration system and mathematical substantiation of the maintenance of functioning of the cogeneration system, was performed. Time constants and coefficients of the mathematical models of dynamics regarding the estimation of a change in the moisture content of the air, temperature of local water, were determined. Functional estimation of a change in the ratio of production and consumption of electric power and heat of the cogeneration system in the range of 0.6294 – 0.6305, moisture content of the air in the range of 12–40%, temperature of local water in the range of 90–79.83°C was obtained. Determining final functional information provides an opportunity to make forestalling decisions to changing the ratio of production and consumption of electric power and heat to maintain the functioning of the cogeneration system.

Keywords: Smart Grid technologies, cogeneration system, drying plant, the heat exchanger for heating air with local water of second circuit of the cogeneration plant

Introduction

Making pellets is the process of converting waste into solid fuel. The fuel pellets obtained during this transformation are used in the process of thermal combustion. Quite often, pellets are produced from such biomass as wood or agricultural waste. Thus, pelleting can be considered as a way of using biomass residues that would otherwise remain unused. Wood pellets are most widely used. However, straw and sunflower husks are also often used as raw materials for pellet production. In the production of the pellet fuel, the costs of timber drying make up to 25% of the total costs. The moisture content should not exceed 10 – 12%, and raw timber, for example, may contain about 50% of water. Moreover, the measurement of the temperature and the air humidity as a drying agent in the drying chamber and the humidity of the dried timber, are not always used appropriately to support timber drying due to the complexity of measurements, which makes it impossible to use the measurements in the coherence to prevent the influence on the change in the parameters of drying for ensuring the continuous production of the pellet fuel.

The most important indicator of raw materials drying is the drying capacity, which must be more than 1 and present the quantitative ratio of the moisture content of raw materials to its balanced moisture content. The balanced moisture content of timber is almost equal to the stable moisture content, which depends on the temperature of the air and its relative humidity. To maintain the quality of drying, it is necessary to coordinate the temperature and the aerodynamic timber drying modes on the basis of the prediction of the air moisture content change in a drying chamber.

On the basis of mathematical and logical modeling as a part of cogeneration system the technology of functioning of the drying installation at the level of deci-sion-making concerning production of pellet fuel is developed. The author, Chaikovskaya, E. (2016) proposes an integrated system for estimating changes in the moisture content of air in the drying chamber, which allows to change the flow of air supplied for heating, based on changes in engine speed of the air fan, when measuring the air temperature at the outlet of the drying chamber. The exact time of shipment of dried wood and supply of fresh raw materials to the drying unit is determined. It is proposed to predict the change of local water temperature in the second circuit of the cogeneration system to heat the air supplied to the drying chamber based on the change of the heat exchange surface of the heat exchanger of the second circuit of local water heating during wood drying, drying of dried material and loading of fresh raw materials. Round-the-clock operation of pellet plants allows to extend the term of operation of cogeneration systems in terms of maintaining the ratio of electricity production and heat for additional energy production with a guaranteed supply of pellet fuel.

Methodological and Mathematical Substantiation

One of the main properties of energy systems is the mandatory exchange of substance, energy and information with the environment. Thus, the cogeneration system on pellet fuel – open integrated dynamic system, the operation of which requires maintaining the ratio of production and consumption of electricity and heat when using the heat of local water of second heating circuit for air heating. The operation of the cogeneration system on pellet fuel can be considered, in this regard, as the reproduction of external and internal influences and changes in initial conditions, such as changes in cogeneration power, changes in air temperature at the outlet of the heat exchanger, changes in flue gases temperature at the inlet of the heat exchanger of second local water heating circuit, return water temperature, etc. The nature of the reaction is determined by the inertia of the devices and the rate of transients, i.e., the dynamic properties.

Dynamic properties are manifested in the operation of the cogeneration system, information about which can be obtained from the dynamic characteristics, such as changes in air humidity in the drying chamber, changes in temperature of local water at the outlet of the second heating circuit. The dynamic characteristics of the cogeneration system with sufficient accuracy for practice can be described by a finite set of parameters for changes in time, and the spatial coordinate that coincides with the direction of flow of the medium, such as changes air humidity in the drying chamber, changes temperature of local water at the outlet of the second heating circuit, etc. Input effects can also be described by a set of parameters, such as changes in the temperature of air at the outlet of the heat exchanger, changes the temperature of the flue gases at the inlet of the heat exchanger, changes in return water temperature. Therefore, the dynamic description of the cogeneration system on pellet fuel most fully and multifacetedly characterizes its operation. Thus, it is possible to determine that the real cogeneration system is a dynamic system, the mathematical model of which reflects the properties of the transformation of influences, i.e., its dynamic properties. Due to the fact that the cogeneration system reflects the dynamic peculiarities due to the nature of reactions to influences, the supporting of the functioning cogeneration system should be part of such a technological system, which is based on a dynamic system.

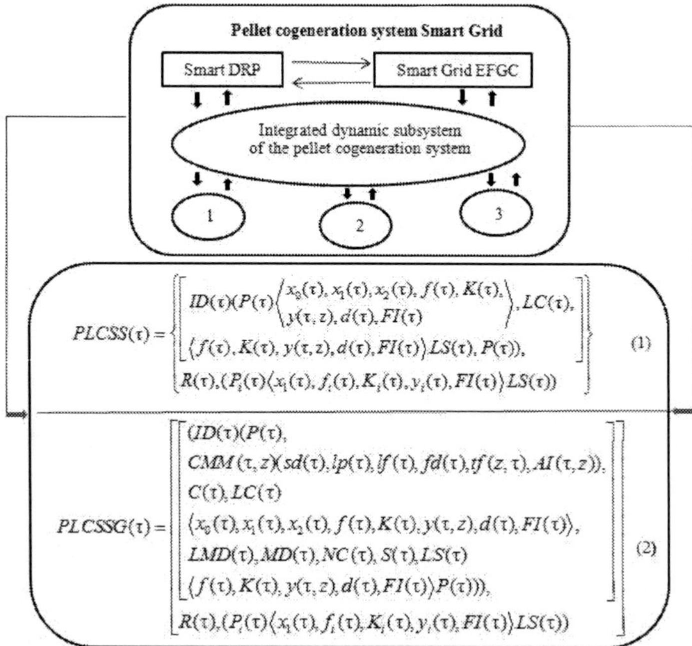

Figure 4.1. Pellet cogeneration system Smart Grid: the architecture: DRP – drying plant; EFGC – engine flue gas circuit; 1– charge unit; 2–discharge unit;3 – unit of evaluation of functional efficiency. Mathematical substantiation of the architecture (1). Mathematical substantiation of maintenance of the operation (2).

Based on the methodological, mathematical, logical substantiation of the technological systems [Chapter 1], energy-saving technology of the drying plant (the author, Chaikovskaya, E. (2016)), the architecture, mathematical substantiation of the architecture (1), mathematical substantiation of maintenance of the operation (2) of the pellet cogeneration system Smart Grid are proposed (Figure 4.1).

It is proposed to include a drying chamber, a heat exchanger for heating air using local water from the second circuit of the cogeneration unit and an air fan. (Figure 4.1). A cogeneration system is a dynamic system, the operation of which is the reproduction of a change in external, internal influences and initial conditions, for example, changes in air temperature at the outlet from the drying chamber, gas temperature in the inlet of heat exchanger in the second circuit of the cogeneration system, return water temperature, etc. (Figure 4.1). That is why, when designing a cogeneration system, underlying which is an integrated dynamic subsystem (Figure 4.1). The integrated dynamic subsystem includes the following components: the electric network, the cogeneration plant, the second circuit as a part of the cogeneration plant, the drying chamber, the heat exchanger for heating air with local water of second circuit of the cogeneration plant, fan. Other units included in the cogeneration system are the units of charge, discharge, and estimation of functional efficiency, which are in harmonized interaction with the integrated dynamic subsystem (Figure 4.1).

The mathematical substantiation of the architecture of the pellet cogeneration system Smart (1), (Figure 4.1), based on the methodology of the mathematical description of dynamics of power systems, the method of the graph of cause-effect relations [Chapter 1] is proposed. (Figure 4. 1).

Where $PLCSS(\tau)$ – pellet cogeneration system Smart; τ – time, s; $ID(\tau)$ – integrated dynamic subsystem (the electric network, the cogeneration unit, the second circuit as a part of the cogeneration plant, the drying chamber, the heat exchanger for heating air with local water of second circuit of the cogeneration plant, air fan; $P(\tau)$ – properties of the components of the cogeneration system; $x(\tau)$ – impacts (change in temperature of air at the outlet from the drying chamber, temperature of flue gas at the inlet to the heat exchanger of second circuit of the cogeneration plant, temperature of return water); $f(\tau)$ – parameters that are measured: (the temperature of air at the outlet from the drying chamber, temperature of flue gas at the inlet to the heat exchanger of second circuit of the cogeneration plant, temperature of return water); $K(\tau)$ – coefficients of mathematical description of the dynamics of air moisture content, the temperature of local water; $y(\tau, z)$ – output parameters (air moisture content, temperature of local water);z – spatial coordinate of the air heat exchanger, local water heat exchanger coincides with the direction of the flow of motion of the medium, m; $d(\tau)$ – dynamic parameters of air moisture content, temperature of local water; $FI(\tau)$ – functional resulting information on decision making; $LC(\tau)$ – logical relations regarding the control of the cogeneration system workability; $LS(\tau)$ – logical relations regarding the identification of the state of the cogeneration system; $R(\tau)$ – logical relations in $PLCSS$ (τ) to confirm the correctness of decisions made from the units of the cogeneration system.. Indices: i – the number of elements of the cogeneration system; 0, 1, 2 – initial stationary mode, external, internal nature of impacts.

The mathematical substantiation of maintenance of the operation of the pellet cogeneration system Smart Grid (2), (Figure 4.1), based on the methodology of the mathematical description of dynamics of power systems, the method of the graph of cause-effect relations [Chapter 1] is proposed. The basis of the proposed rationale is the mathematical description of the architecture of the pellet cogeneration system Smart (1), (Figure 4. 1). Prediction of a change in the air moisture content, temperature of local water of the second circuit of the cogeneration plant makes it possible to make forestalling decisions on a change of the speed of the air fan motor in relation to the change in the flow of air supplied for heating and the number of plates of the heat exchanger of the second circuit of the cogeneration unit The temperature of air at the outlet from the drying chamber, temperature of flue gases at the inlet to the heat exchanger of second circuit of the cogeneration plant, temperature of return water are measured. Mathematical substantiation of Smart Grid maintenance of the operation of the pellet cogeneration system (2) is proposed (Figure 4. 1).

Where $PLCSSG(\tau)$ – pellet cogeneration system Smart Grid; τ – time, s; $ID(\tau)$ – integrated dynamic subsystem (the electric network, the cogeneration unit, the second circuit as a part of the cogeneration plant, the drying chamber, the heat exchanger for heating air with local water of second circuit of the cogeneration plant, air fan; $P(\tau)$ – the properties of the elements of the integrated dynamic subsystem, units of the cogeneration system; $CMM(\tau, z)$ – complex mathematical modeling of the dynamics of changes in the air moisture content, the temperature of local water of second circuit of the cogeneration plant; $sd(\tau)$ – the input data (productivity of the drying plant and the type of the cogeneration system and its power; the temperature of air at the inlet of the heat exchanger and at the outlet of the heat exchanger, the temperature of gas at the inlet of the heat exchanger and at the outlet of the heat exchanger, type of the fan and its power; $lp(\tau)$ – the boundary change in parameters (the temperature of air at the inlet of the heat exchanger, equilibrium moisture content, the temperature of gas at the inlet of the heat exchanger; $lf(\tau)$ – the levels of operation of the drying plant, cogeneration system; $fd(\tau)$ – the obtained parameters (parameters of heat exchange in the heat exchanger of the heating air, second circuit of the heating local water, time constants and coefficients of the mathematical model of dynamics of a change in the air moisture content, the temperature of local water; $tf(\tau,z)$ – the transfer function of predicted parameters – the air moisture content, the temperature of local water of second circuit of the cogeneration plant; $AI(\tau,z)$ – the standard information regarding the evaluation of the maximum admissible change in the air moisture content, the temperature of local water; $C(\tau)$ – the control of workability of the cogeneration system; $LC(\tau)$ – the logical relations of the control of the cogeneration system workability; $x(\tau)$ – impacts (change in temperature of air at the outlet from the drying chamber, temperature of gas at the inlet at the heat exchanger, temperature of return water)); $f(\tau)$ – the measured parameters:(temperature of air at the outlet from the drying chamber, temperature of gas at the inlet at the heat exchanger, temperature of return water); $K(\tau)$ – the coefficients of the mathematical description of the dynamics of a change in the air moisture content, of a change in the temperature of local water; $y(\tau, z)$ – the output parameters (the air moisture content, the temperature of local water); z – the coordinate of the length of the heat exchanger of the air, the heat exchanger of the second circuit of the cogeneration system, m; $d(\tau)$ – the dynamic parameters of estimation of a change in the air moisture content, of a change in the temperature of local water; $FI(\tau)$ – functional resulting information on decision making; $LMD(\tau)$ – the logical relations of decision making; $MD(\tau)$ – decision making; $NC(\tau)$ – the new conditions of the cogeneration system operation; $S(\tau)$ – the identification of the state of the cogeneration system; $LS(\tau)$ – the logical relations of identification of the state of the cogeneration system; $R(\tau)$ – the logical relations between the dynamic subsystem and units of charge, discharge, functional estimation of efficiency that belong to the cogeneration system. Indices: i – the number of elements of $PLCSSG(\tau)$ (τ); 0, 1, 2 – the initial, external, and internal character of influences.

Mathematical substantiation of the architecture of the cogeneration system (1) and mathematical description (2) makes it possible to maintain the operation of the cogeneration system using the following actions.

- Workability control ($C(\tau)$) of the dynamic subsystem based on complex mathematical ($CMM(\tau, z)$) and logical ($LC(\tau)$) modeling regarding obtaining standard ($AI(\tau,z)$) estimate of a change in the air moisture content, the temperature of local water;

- Workability control ($C(\tau)$) of the dynamic system based on complex mathematical ($CMM(\tau, z)$) and logical ($LC(\tau)$) modeling regarding the obtaining functional ($FI\ (\tau)$) estimate of a change in the air moisture content, of a change in the temperature of local water;
- Decision making ($MD(\tau)$) with the use of the resulting functional information ($FI\ (\tau)$), obtained based on logical modeling ($LMD(\tau)$); decision making to maintain the ratio of production and consumption of electric power and heat with the use of the functional assessment of a change in the air moisture content, of a change in the temperature of local water;
- Identification ($S(\tau)$) of the new conditions of functioning of the cogeneration system ($NC(\tau)$) based on logical modeling ($LS(\tau)$) as a part of the dynamic subsystem and confirmation of new operating conditions based on logical modeling ($R(\tau)$) from the units of the cogeneration system.

Complex Mathematical and Logical Modeling of the Pellet Cogeneration System Smart Grid

According to formulas (1), (2), the prediction of a change in the air moisture content is proposed. The temperature of air at the outlet from the drying chamber is measured. The transfer function along the channel "air moisture content – air consumption" is obtained. A change in the air moisture content is estimated both over time and along the spatial coordinate of the heat exchanger coincides with the direction of the flow of motion of the medium. The transfer function along the channel "air moisture content – air consumption," which was obtained as a result of solving a system of nonlinear differential equations, is presented as follows:

$$W_{w-Gair_1} = \frac{K_{\text{air}}(\beta - 1)}{K_{\text{w}}\beta\gamma}\left(1 - e^{-\gamma\xi}\right), \quad (3)$$

where

$$K_{air} = \frac{m(\theta_0 - \sigma_0)}{G_{air0}}; \quad \beta = T_m S + \varepsilon^* + 1; \quad T_m = \frac{g_m C_m}{\alpha_{air0} h_{air0}};$$

$$\varepsilon^* = \varepsilon(1 - L_{lw}^*); \; \varepsilon = \frac{\alpha_{lw0} h_{lw0}}{\alpha_{air0} h_{air0}}; \; L_{lw}^* = \frac{1}{L_{lw} + 1};$$

$$L_{lw} = \frac{G_{lw} C_{lw}}{\alpha_{lw0} h_{lw0}}; \; K_w = \frac{\partial i}{\partial w} / \frac{\partial i}{\partial t}; \; \gamma = T_{air} S;$$

$$T_{air} = \frac{g_{air} C_{air}}{\alpha_{air0} h_{air0}}; \; \xi = \frac{z}{L_{air}}; \; L_{air} = \frac{G_{air} C_{air}}{\alpha_{air0} h_{air0}},$$

where C is the specific thermal capacity, kJ/(kg·K); α is the heat transfer factor, kW/(m^2·K); G is the consumption of substance, kg/s; g is the specific weight of a substance, kg/m; h is the specific surface, m^2/m; t, σ, θ are the temperature of the air, local water and of the separating wall, respectively, K; i is enthalpy of the working fluid, kJ/kg; z is spatial coordinate of the heat exchanger coincides with the direction of the flow of motion of the medium, m; $T_{air,}$, T_m are the time constants that characterize the thermal accumulating capacity of air, metal, s; m is the indicator of the dependence of heat transfer factor on consumption; τ is the time, s; S is the Laplace parameter; $S = \omega j$; ω is the frequency, 1/s. Indices: 0 – initial stationary mode; 1 – inlet to the heat exchanger; air – internal flow – air; lw – external flow –local water; w– air moisture content; m – metal wall.

Transfer function along the channel: "air moisture content – air consumption" was obtained based on the solution of a system of nonlinear differential equations using the Laplace transform tool. The system of differential equations includes the equation of state as the estimation of the physical model of the drying chamber. The system of differential equations also includes the equation of energy of transmitting and receiving media – local water and air, respectively, and the equation of thermal balance for the wall of the heat exchanger of the air. The equation of the energy of the receiving medium is developed with the representation of a change in air moisture content both in time and along the spatial coordinate, which coincides with the direction of the motion flow of the medium. The equation of the energy of the receiving medium includes the K_{air} coefficient, which assesses a change in the temperature of air at the outlet of the drying chamber that is measured. The equation of the energy of the transmitting medium includes the K_{lw} coefficient, which assesses a change in the temperature of local water at the inlet of the heat exchanger that is measured.

A real part of the transfer function was separated:

$$O(\omega) = (C_1 L_1 - D_1 M_1)(K_{air} / K_w) \tag{4}$$

The K_{air} factor includes the temperature of the separating wall θ:

$$\theta = \left(\alpha_{air}\left(\sigma_1 + \sigma_2\right)/2\right) + \left(A\left(t_1 + t_2\right)/2\right)/\left(\alpha_{air} + A\right), \tag{5}$$

where σ_1, σ_2 are the temperatures of local water at the inlet and at the outlet of the heat exchanger, K, respectively; t_1, t_2 are the temperatures of air at the inlet and at the outlet of the heat exchanger, K, respectively; α is the heat transfer factor, kW/(m$^2\cdot$K). Index air – internal flow: air.

$$A = 1/\left(\delta_m / \lambda_m + 1/\alpha_{lw}\right), \tag{6}$$

where δ is the thickness of a wall of the heat exchanger, m; λ is the thermal conductivity of the metal wall of the heat exchanger, kW/(m$\cdot$K). Indices: lw – external flow: local water; m – metal wall of a heat exchanger.

To use the real part $O(\omega)$, the following factors were obtained:

$$A_1 = \varepsilon * - T_{air} T_m \omega^2 \tag{7}$$

$$A_2 = - T_{air} T_m \omega^2 \tag{8}$$

$$B_1 = T_m \omega \tag{9}$$

$$B_2 = T_{air} \omega (\varepsilon + 1) \tag{10}$$

$$C_1 = \frac{A_1 A_2 + B_1 B_2}{A_2^{\,2} + B_2^{\,2}} \tag{11}$$

$$D_1 = \frac{A_2B_1 - A_1B_2}{A_2^{\ 2} + B_2^{\ 2}} \tag{12}$$

$$L_1 = 1 - \cos(-\xi T_{\mathrm{air}}\omega) \tag{13}$$

$$M_1 = 1 - \sin(-\xi T_{air}\omega) \tag{14}$$

The transfer function (3), which was obtained based on the use of the operator method of solving the system of nonlinear differential equations, retains the Laplace transform parameter – S ($S = \omega j$), where ω is the frequency, 1/s. To switch from the frequency area to the time area, a real part (4), obtained as a result of the mathematical treatment of transfer function, was separated. It is this part that is included in the integrals (15), which makes it possible to obtain dynamic characteristics of a change the air moisture content using the inverse Fourier transform:

$$w(\tau, z) = \frac{1}{2\pi}\int_0^{\infty} O(\omega)\sin(\tau\omega/\omega)d\omega \quad , \tag{15}$$

where w is air moisture content, %

According to formulas (1), (2), the prediction of a change in the temperature of local water is proposed. In the second circuit of the cogeneration system, the following parameters are measured: the temperature of the flue gases at the inlet of the heat exchanger and the return water temperature. The transfer function along the channel "temperature of local water – temperature of flue gases" is obtained. A change in the local water temperature is estimated both over time and along the spatial coordinate of the heat exchanger coincides with the direction of the flow of motion of the medium.

The transfer function along the channel " temperature of local water – temperature of flue gases," which is obtained as a result of solving a system of nonlinear differential equations, is presented as follows:

$$W_{t-\vartheta_1} = \frac{K_g \varepsilon \left(1 - L_g^*\right)}{\left(T_{lw} S + 1\right)\beta - 1}\left(1 - e^{-\gamma\xi}\right), \tag{16}$$

where

$$K_g = \frac{m\left(\theta_0 - \sigma_0\right)}{G_{g0}}; \quad \varepsilon = \frac{\alpha_{g0} h_{g0}}{\alpha_{lw0} h_{lw0}}; \quad L_g = \frac{G_g C_g}{\alpha_{g0} h_{g0}};$$

$$T_{lw} = \frac{g_{lw} C_{lw}}{\alpha_{lw0} h_{lw0}}; \quad \beta = T_m S + \varepsilon^* + 1; \quad T_m = \frac{g_m C_m}{\alpha_{lw0} h_{lw0}}; \quad \varepsilon^* = \varepsilon\left(1 - L_g^*\right);$$

$$\gamma = \frac{\left(T_{lw} S + 1\right)\beta - 1}{\beta}; \quad \xi = \frac{z}{L_{lw}}; \quad L_{lw} = \frac{G_{lw} C_{lw}}{\alpha_{lw0} h_{lw0}},$$

where C is the specific thermal capacity, kJ/(kg·K); α is the heat transfer factor, kW/(m^2·K); G is the consumption of substance, kg/s; g is the specific weight of a substance, kg/m; h is the specific surface, m^2/m; t, σ, θ are the temperature of the local water, gas and of the separating wall, respectively, K; z is spatial coordinate of the heat exchanger coincides with the direction of the flow of motion of the medium, m; T_{lw}, T_m are the time constants that characterize the thermal accumulating capacity of local water, metal, s; m is the indicator of the dependence of heat transfer factor on consumption; τ is the time, s; S is the Laplace parameter; $S = \omega j$; ω is the frequency, 1/s.; Indices: 0 – initial stationary mode; 1 – inlet to the heat exchanger; lw – internal flow – local water; g – external flow – flue gases; m – metal wall.

Transfer function along the channel: "temperature of local water – temperature of flue gases" was obtained based on the solution of a system of nonlinear differential equations using the Laplace transform tool. The system of differential equations includes the equation of state as the estimation of the physical model of the cogeneration system. The system of differential equations also includes the equation of energy of transmitting and receiving media – flue gases and local water, respectively, and the equation of thermal balance for the wall of the heat exchanger of the second circuit. The equation of the energy of the receiving medium is developed with the representation of a change in local water both in time and along the spatial coordinate, which coincides with the direction of the motion flow of the medium. The equation of the energy of the transmitting medium includes the coefficient $K_{g.}$ which assesses a change in the temperature of flue gases at the inlet of the heat exchanger and the temperature of return water are measured.

A real part of the transfer function was separated:

$$O(\omega) = \frac{(L_1 A_1) + (M_1 B_1)(1 - L_3^*)}{(A_1^2 + B_1^2)}. \tag{17}$$

The K_g factor includes the temperature of the separating wall θ:

$$\theta = (\alpha_{lw}(\sigma_1 + \sigma_2)/2) + (A(t_1 + t_2)/2)/(\alpha_{lw} + A), \tag{18}$$

where σ_1, σ_2 are the temperatures of flue gases at the inlet and at the outlet of the heat exchanger, K, respectively; t_1, t_2 are the temperatures local water at the inlet and at the outlet of the heat exchanger, K, respectively; α is the heat transfer factor, kW/(m^2·K). Index lw – internal flow: local water.

$$A = 1/(\delta_m/\lambda_m + 1/\alpha_g), \tag{19}$$

where δ is the thickness of a wall of the heat exchanger, m; λ is the thermal conductivity of the metal wall of the heat exchanger, kW/(m·K). Indices: g – external flow: flue gases; *m* – metal wall of a heat exchanger.

To use the real part $O(\omega)$, the following factors were obtained:

$$A_1 = \varepsilon^* - T_{lw} T_m \omega^2; \tag{20}$$

$$A_2 = \varepsilon^* + 1; \tag{21}$$

$$B_1 = T_{lw} \varepsilon^* \omega + T_{lw} \omega + T_m \omega; \tag{22}$$

$$B_2 = T_m \omega; \tag{23}$$

$$C_1 = \frac{A_1 A_2 + B_1 B_2}{A_2^2 + B_2^2}; \tag{24}$$

$$D_1 = \frac{A_2 B_1 - A_1 B_2}{A_2^{\ 2} + B_2^{\ 2}}; \tag{25}$$

$$L_1 = 1 - e^{-\zeta C_1} \cos\left(-\xi D_1\right); \tag{26}$$

$$M_1 = -e^{-\zeta C_1} \sin\left(-\xi D_1\right). \tag{27}$$

The transfer function (16), which was obtained based on the use of the operator method of solving the system of nonlinear differential equations, retains the Laplace transform parameter – S (S=ωj ω is the frequency, 1/s. To switch from the frequency area to the time area, a real part (17), obtained as a result of the mathematical treatment of transfer function, was separated. It is this part that is included in the integral (28), which makes it possible to obtain dynamic characteristics of a change in power factor of the cogeneration system, the temperature of local water using the inverse Fourier transform.

$$t\left(\tau, z\right) = \frac{1}{2\pi} \int_0^{\infty} K_g O\left(\omega\right) \sin\left(\tau\omega/\omega\right) d\omega, \tag{28}$$

where t is the temperature of local water, K.

Thus, for obtaining the reference estimation of change air moisture content and temperature of local water a block diagram (Figure. 4.2) is proposed using, for example, the initial data of the pellet cogeneration system of the *GTK* 35M type total power 112 kW, electrical power 35 kW, heat power 60 kW.

The cogeneration system includes a drying plant with a capacity of Q_{dr}= 680 kg/day of wood. Consumption of air for drying – 1.8 kg/s and consumption of local water for air heating – 0.43 kg/s. Equilibrium air humidity is 20%. The following levels of operation of the drying plant have been established for the change in the temperature of the air at the inlet to the heat exchanger and at the outlet from the heat exchanger: – first level: 55–85°C; second level: 55.6–80.8°C; third level: 51.3–77°C. The set levels of operation correspond to the change in equilibrium humidity – 20%, 15%, 12% and air consumption 1.8 kg/s, 1.68 kg/s, 1.65 kg/s, respectively, to reduce the humidity of raw materials from 40% to 12%. The following levels of operation of the cogeneration system have been established for the change in the temperature of the outgoing gases at the inlet to the heat exchanger and at the outlet from the heat exchanger: – first level: 140–65°C; second level: 130–65°C; third level: 120–65°C for heating 0.43 kg/s of local water from 60°C to 90°C. They correspond to changes in the number of plates of the heat exchanger: 36, 44, 52 and to changes in outgoing gases consumption: 0.69 kg/s, 0.79 kg/s, 0.94 kg/s, respectively.

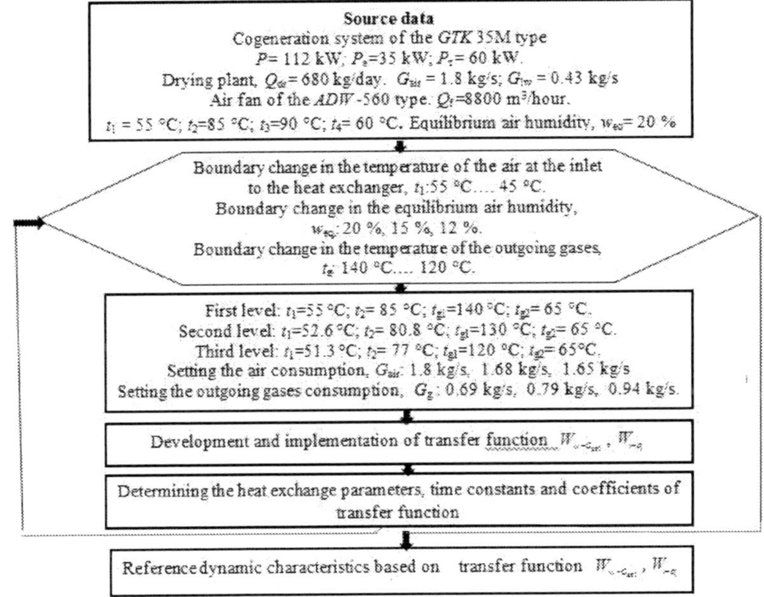

Figure 4.2. Block diagram of complex mathematical modeling of the cogeneration pellet system: P, P_e, P_t are the nominal, electric, thermal power of the cogeneration system, respectively, kW; Q_{dr} is productivity of drying plant, kg/day; G_{air}, G_{lw} are the consumption air, local water, respectively, kg/s; Q_f is productivity of axial fan, m^3/hour; t_1, t_2, are the temperature of air at the inlet of the heat exchanger and at the outlet of the heat exchanger, respectively, °C; t_3, t_4 are the temperature of local water at the inlet of the heat exchanger and at the outlet of the heat exchanger, respectively, °C; t_{g1}, t_{g2} are the temperature of outgoing gases at the inlet of the heat exchanger and at the outlet of the heat exchanger, respectively, °C; w_{eq} is equilibrium air humidity, %.

According to Formulas (1) - (3), (16) and proposed block diagram (Figure 4.2), the results of complex mathematical modeling of the pellet cogeneration system are presented (Tables 4.1–4.4).

As presented in Tables 4.2, 4.4, the time constants and coefficients that are part of the mathematical model of dynamics (3, 16) are obtained on the basis of parameters as part of complex mathematical modeling (Tables 4.1, 4.3).

Table 4.1. Heat transfer parameters as part of complex mathematical modeling of air heat exchanger

Levels of operation	Parameter		
	α_{lw}, kW/(m²·K)	α_{air}, kW/(m²·K)	k, kW/(m²·K)
First level	3.051	0.254	0.233
Second level	2,913	0.243	0.223
Third level	2.878	0.240	0.220

Note: α_{lw} – coefficient of convective heat transfer from the local water to the heat exchanger wall, kW/(m²·K); α_{air} – coefficient of convective heat transfer from the heat exchanger wall to air, kW/(m²·K); k – heat transfer coefficient, kW/(m²·K).

Table 4.2. Time constants and coefficients of the mathematical model of the dynamics of the moisture content of the air of the drying plant

Levels of operation	T_{air}, s	T_m, s	ε	ζ	L_{lw}, m	L_{air}, m	L_{lw}^*	K_w
First level	0.0048	0.6149	14.547	0.146	33.26	483.89	0.029	-0. 9646
Second level	0.0050	0.6440	14.547	0.136	32.51	472.99	0.030	-1.2848
Third level	0.0051	0.6518	14.547	0.134	32.32	470.9	0.030	-1.6060

Table 4.3. Heat transfer parameters as part of complex mathematical modeling of local water heat exchanger

Levels of operation	Parameter		
	α_g, kW/(m²·K)	α_{lw}, kW/(m²·K)	k, kW/(m²·K)
First level	3.596	1.028	0.746
Second level	3.133	0,961	0.732
Third level	2.755	0.922	0.698

Note: α_g – coefficient of convective heat transfer from the gas to the heat exchanger wall, kW/(m²·K); α_{lw} – coefficient of convective heat transfer from the heat exchanger wall to local water, kW/(m²·K); k – heat transfer coefficient, kW/(m²·K).

Table 4.4. Time constants and coefficients of the mathematical model of the dynamics of the local temperature of the second circuit of the cogeneration system

Levels of operation	$T_{lw,\ s}$	$T_{m,\ s}$	$L_{g,\ m}$	$L_{lw,\ m}$	$L_{g.}^*$	ζ	ε
First level	4.67	1.52	11.4	114.9	0.08	0.67	4.07
Second level	4.79	1.62	15.1	123.0	0.06	0.76	3.79

Third level	4.99	1.69	20.3	128.2	0.05	0.78	3.47

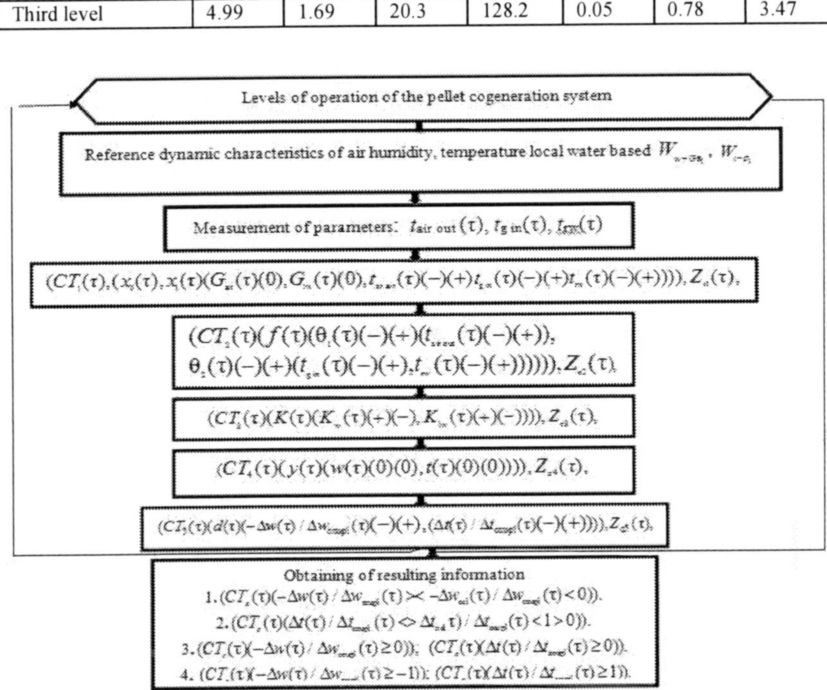

Figure 4.3. Block diagram of control of efficiency of the cogeneration system: $t_{air\ out}(\tau)$, $t_{g\ in}(\tau)$, $t_{rw}(\tau)$ are the temperatures of air at the outlet of the heat exchanger, flue gases at the inlet of the heat exchanger of the second heating circuit, the temperature of return water, respectively, °C; θ_1, θ_2 are the temperatures of the separating wall of the air heat exchanger, of the heat exchanger of the second heating circuit, respectively, K; *CT* is the event control; *Z* is the logical relations; *d* is the dynamic parameters; *x* is the influences; *f* is the diagnosed parameters; *y* is the output parameters; *K* is the mathematic description factors; *w* is air humidity, %; *t is* temperatures of local water, K; ι is time. Indices: *c* is efficiency control; air is air; g is flue gases; rw is return water; ccupl, ccl are constant, calculated value of the parameter of the upper level of functioning, level of functioning, respectively; 0, 1, 2 are the initial stationary mode, external, internal parameters; 3 is the coefficients of dynamics equations; 4 is the essential diagnosed parameters; 5 is dynamic parameters.

Based on the proposed mathematical substantiation Smart Grid maintenance of functioning of the cogeneration system (1) to (3), (16) for obtaining functional estimation of a change in the air moisture content and temperature of local water the block diagram for the control of efficiency of the cogeneration on pellet fuel (Figure 4.3) is developed.

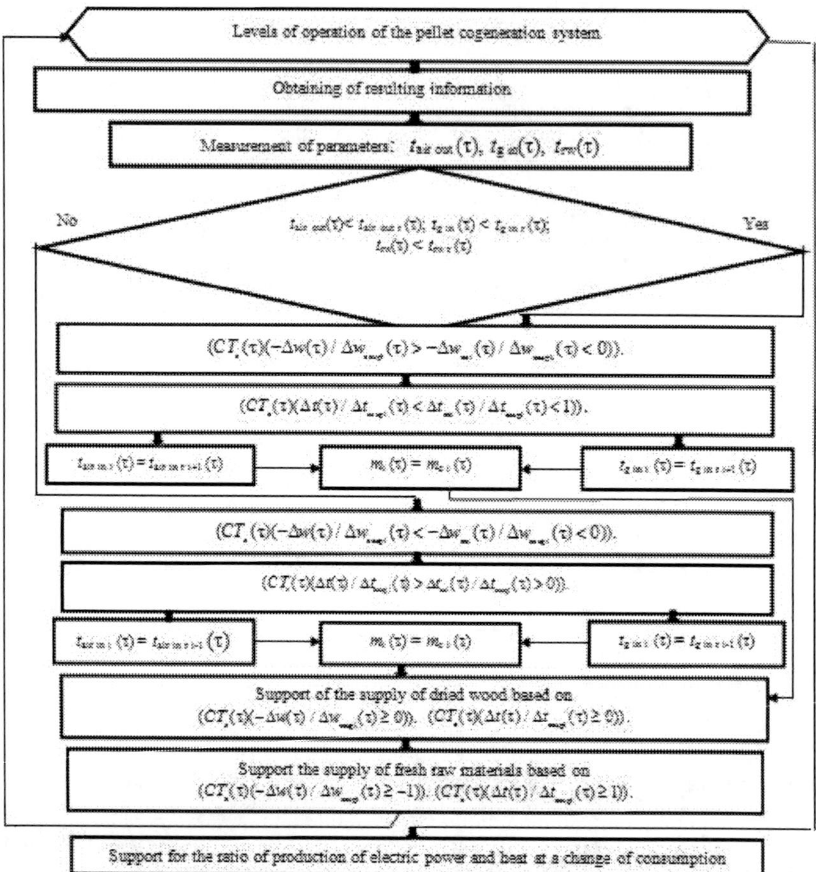

Figure 4.4. Block diagram of maintenance of functioning of the cogeneration system: CT_c is the efficiency control; $t_{\text{air out}}(\tau)$, $t_{\text{g in}}(\tau)$, $t_{\text{rw}}(\tau)$ are the temperatures of air at the outlet of the heat exchanger, flue gases at the inlet of the heat exchanger of the second heating circuit, the temperature of return water, respectively, K; w is the air humidity, %; t is the temperatures of local water of the second heating circuit, K; ι is the time; m is the ratio of electricity and heat production. Indices: i is the number of operation levels; r is the reference value of the parameter; in, outlet are the inlet of the heat exchanger, the outlet of the heat exchanger, respectively; ccupl, ccl, are the constant calculation value of the parameter of the upper level of functioning, level of functioning, respectively.

Control of serviceability of the cogeneration system (Figure 4.3) enables obtaining the resulting information on decision-making about the maintenance of the functioning of the cogeneration system.

Smart Grid System of Maintaining the Operation of the Cogeneration Pellet System at the Decision-Making Level

Based on the proposed mathematical substantiation (1) to (3), (16), the block diagram of the maintenance of the functioning of the cogeneration pellet system (Figure 4.4) are presented.

Results and Discussion

The comprehensive integrated system of maintenance of operation of the cogeneration system is developed. (Table 4.5). There is a continuous measurement of the temperature of air at the outlet from the drying chamber. In the second heating circuit the temperature of flue gases at the inlet of the heat exchanger and the temperature of return water are measured.

The moisture content in the drying chamber at a definite time is determined as follows:

$$w_{i+1}(\tau) = w_i + \\ + \begin{pmatrix} \Delta w_{i+1}(\tau) / \Delta w_{\text{ccupl.}}(\tau) - \\ -\Delta w_i(\tau) / \Delta w_{\text{ccupl.}}(\tau) \end{pmatrix} (w_2 - w_1), \qquad (29)$$

where w is the moisture content of the air in the drying chamber, %; w_1, w_2 are the moisture content of the air at the inlet to the heat exchanger of the air heating and at the outlet of the heat exchanger, %, respectively; i is the number of levels of timber drying; τ is the time, s. Index: ccupl – the constant calculation value of the parameter of the upper level of functioning

A temperature of the local water in the established period is determined as follows (Table 4.5):

$$t_{lwi+1}(\tau) = t_{lwi} - \\ - \begin{pmatrix} \Delta t_{lwi+1}(\tau) / \Delta t_{\text{ccup.}}(\tau) - \\ -\Delta t_{lwi}(\tau) / \Delta t_{\text{ccup}}(\tau) \end{pmatrix} (t_2 - t_1), \qquad (30)$$

where t_{lw} is temperature of local water, °C ; t_1, t_2 are the initial, final values of temperature of local water, °C; τ is the time, s. Index: ccup is the constant, calculated value of the parameter of the upper level of functioning; i is the number of levels of the cogeneration system.

Thus, for example, in 4400 s after loading raw wood to the drying chamber, it was established that the aerodynamic mode of the drying chamber regarding the air supply of 1.8 kg/s into the heat exchanger does not correspond to the reduction of the air temperature at the outlet of the drying chamber to the level of 52°C and requires to make a decision on the change of air consumption to the level of 1.68 kg/s regarding the coordination of aerodynamic and thermal regimes of timber drying. The absolute value of the air humidity in the drying chamber at this period of time was determined as follows:

$$18.87\% = 17.46\% + ((0.8048 - 0.7544)(40\% - 12\%)).$$

The absolute value of the temperature of local water (Table 4.5) with the use of formula (30) in the period $44 \cdot 10^2$ s (1.8 hours) is:

$$87.25°C = 87.84°C - (0.9246 – 0.9049)(90 – 60°C).$$

For further timber drying, it is necessary to control the air temperature at the outlet of the drying chamber by reducing the power of the air fan regarding the air supply of 1.68 kg/s, which in 6600 s, makes 51°C, and to make the further timber drying to the air humidity in the drying chamber, for example, of 24.07%, calculated as follows:

$$24.07\% = 18.87\% + ((0.7532 - 0.5674)(40\% - 12\%)).$$

And if the air temperature at the outlet of the drying chamber decreased to 44°C, which corresponds to the increase in the air humidity in the drying chamber up to 40%, it is necessary to unload the dried timber to the pellet plant and load raw materials, ensuring setting of the thermal and aerodynamic parameters of the drying plant of the first level of functioning.

Forecasting the change in the local water temperature allows making advance decisions on changing the heat exchange surface of the heat exchanger of the second circuit of the cogeneration system during the wood drying period and supporting the unloading of dried wood and the loading of fresh material. So, for example, the local water temperature change support level 1 (Figure 4.5) provides heating of 1.8 kg/s of air during the time from loading fresh wood to air heating to 82°C.

Table 4.5. Integrated Smart Grid System of harmonization of production and consumption of electric power and heat of the cogeneration pellet system

Time, τ, 100 s	Changing the parameters of the technological process	$\Delta w(\tau)/\Delta w_1(\tau)$	$w(\tau)$, %	$\Delta t_{lw}(\tau)/\Delta t_1(\tau)$	$t_{lw}(\tau)$, °C
11	Loading raw timber. Making decision on the air supply of 1.8 kg/s t_1= 85°C; t_2= 55°C; t_3 =140°C; t_4=65°C; $t_{rw.}$=60°C; n= 36 pieces; Gg=0.69 kg/s; Q_{T1}= 54.4 kW; Q_{T2}=55.6 kW; P_e=34.24 kW; m=0.6294	-1	12	1	90
22	Air supply of 1.8 kg/s. Timber drying: t_1=84°C; t_2=54°C; t_3 =139°C; t_4=65°C; $t_{rw.}$=60°C; n= 36 pieces; Gg=0,69 kg/s; Q_{T1}= 54.4 kW Q_{T2}=54.8 kW; P_e=34.74 kW; m = 0.6386	-0,9107	14,5	0.9533	88.56
33	Air supply of 1.8 kg/s. Timber drying: t_1=82°C; t_2=53°C; t_3 =138°C; t_4=65°C; $t_{rw.}$=60°C; n= 36 pieces; Gg=0,69 kg/s; Q_{T1}= 52.6 kW; Q_{T2}=54.1 kW; P_e=34.03 kW; m=0.6469	-0,8048	17,46	0.9246	87.84
44	Making a decision the air supply of 1.68 kg/s: t_1= 80.5°C; t_2= 52°C; t_3 =130°C; t_4=65°C; $t_{rw.}$= 56°C n= 44 pieces; Gg=0,79 kg/s; Q_{T1}= 48.2 kW; Q_{T2}=55.1 kW; P_e=30.62 kW; m=0.6353	-0,7544	18.87	0.9039	87.25
55	Identification of the new conditions of functioning: air supply of 1.68 kg/s: t_1= 80.5°C; t_2= 52°C; t_3 =130°C; t_4=65°C; $t_{rw.}$= 56°C; n= 44 pieces; Gg=0.79 kg/s; Q_{T1}= 48.2 kW; Q_{T2}=55.1 kW; P_e=30.62 kW; m=0.6353	-0,7532	18,87	0.9039	87.22
66	Air supply of 1,68 kg/s. Timber drying: t_1= 76,6°C; t_2= 51°C; t_3 =129°C; t_4=65°C; $t_{rw.}$= 55°C; n= 44 pieces; Gg=0.79 kg/s; Q_{T1}= 43.3 kW; Q_{T2}=54.3 kW; P_e=27.91 kW m=0.6446	-0,5674	24,07	0..8586	85.86
77	Making decision regarding the air supply of 1.65 kg/s: t_1= 76. 6°C; t_2= 51°C; t_3 =120°C; t_4=65°C; $t_{rw.}$= 55°C; n= 52 pieces; Gg=0,94 kg/s; Q_{T1}= 42.5 kW; Q_{T2}=55.5 kW; P_e=26.8 кВт; m=0,6312	-0,5778	24,36	0.7343	82.13

Table 4.5. (Continued)

Time, τ, 100 s	Changing the parameters of the technological process	$\Delta w(\tau)/\Delta w_1(\tau)$	$w(\tau)$, %	$\Delta t_{lw}(\tau)/\Delta t_1(\tau)$	$t_{lw}(\tau)$, °C
88	Identification of the new conditions of functioning: air supply of 1,65 kg/s: t_1 = 76,6°C; t_2= 51°C; t_3 =120°C; t_4=65°C; $t_{rw.}$= 55°C; n= 52 pieces; Gg=0,94 kg/s; Q_{T1}= 42.5 kW; Q_{T2}=55.5kW; P_e=26.8 kW; m=0.6312	-0,5778	24,36	0.7343	82.13
99	Air supply of 1,65 kg/s. Timber drying: t_1 = 76°C; t_2= 50°C; t_3 =120°C; t_4=65°C; $t_{rw.}$= 55°C; n= 52 pieces; Gg=0,94 kg/s; Q_{T1}= 43.2 kW; Q_{T2}=55.5 kW; P_e=27.2 kW; m=0.6312.	-0,4801	27,10	0.7153	81.56
110	Air supply of 1,65 kg/s. Timber drying: t_1 = 75°C; t_2= 48°C; t_3 =120°C; t_4=65°C; $t_{rw.}$= 55°C; n= 52 pieces; Gg=0,94 kg/s; Q_{T1}= 44.9 kW; Q_{T2}=55.5 kW; P_e=28.3 kW; m=0.6300	-0,3021	32,08	0.6960	80.98
121	Air supply of 1,65 kg/s. Timber drying: t_1 = 75°C; t_2= 45°C; t_3 =120°C; t_4=65°C; $t_{rw.}$= 55°C; n= 52 pieces; Gg=0,94 kg/s; Q_{T1}= 49.8 kW; Q_{T2}=55.5 kW; P_e=31.4 kW; m=0.6305	-0,1277	36,96	0.6770	80.41
132	Unloading dried timber for production of the pellet fuel: t_1 = 74°C; t_2= 44°C; t_3 =120°C; t_4=65°C; $t_{rw.}$= 55°C; n= 52 pieces; Gg=0,94 kg/s; Q_{T1}= 49.8 kW; Q_{T2}=55.5 kW; P_e=31.4 kW; m=0.6305	-0,0190	40	0.6577	79.83

Note: t_1, t_2 are the air temperature at the inlet of the drying chamber and the outlet of the drying chamber, respectively, °C; t_3, t_4 are the fuel gases temperature at the inlet to the heat exchanger and the outlet of the heat exchanger, respectively, °C; t_{rw} is the temperature of the return water, °C; Q_{T1}, Q_{T2} are the thermal power of the air heat exchanger, second heat exchanger, respectively, kW; G is fuel gases consumption, kg/s; n is the number of plates of the second circuit. respectively; P_T, P_e are the active thermal, electric power of the cogeneration system, respectively, kW; m is the ratio of production and consumption of electric power and heat; w is the moisture content of the air in the drying chamber, %; τ is time, s. Index: 1 is constant, calculated value of the parameter of the upper level of functioning.

Making a decision to reduce the rotation frequency of the air fan motor in relation to the change in the flow of air supplied for heating from 1.8 kg/s to 1.68 kg/s (Table 4.5) corresponds to a decrease in the temperature of the gases at the entrance to the heat exchanger of the second circuit of the cogeneration system from 140°C to 134°C and the return water temperature decrease from 60°C to 56°C. At this time, a decision was made to change the number of heat exchanger plates from 36 to 44 in order to enter the tolerance of functioning level 2 (Figure 4.5) of the cogeneration system to maintain the temperature of the local water supplied for air heating.

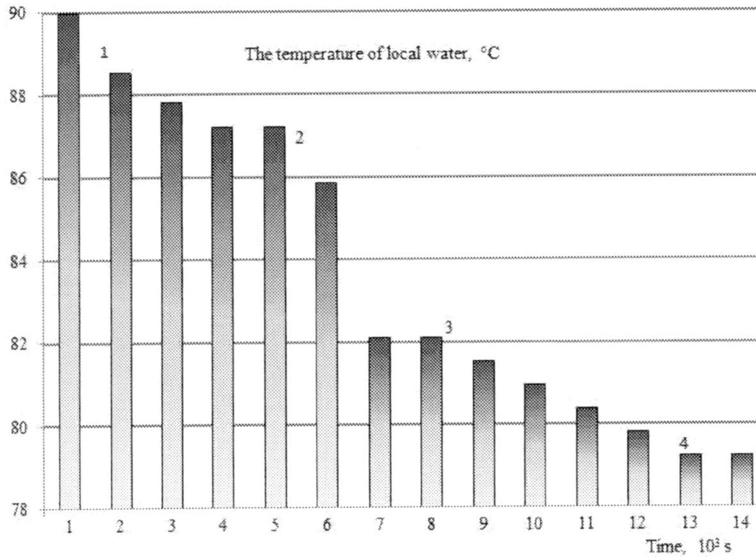

Figure 4.5. Supporting the ratio of electricity and heat production in the cogeneration system: 1– the supply of fresh wood; 2,3– support of the wood drying; 4– the supply of lowding dried wood.

With a further reduction in the frequency of rotation of the electric motor of the air fan during the wood drying period and the corresponding decision-making to change the air consumption for heating (Table 4.5), the obtained integrated system of changing the temperature of the local water (Figure 4.5) allows, on the basis of decisions to change the number of heat exchanger plates, to perform change of functioning level 2 to functioning level 3. In the time period of 13200 s, at the temperature of the air supplied for heating, at the level of 74°C, the decision to ship dried wood corresponds to the change in the temperature of gases at the entrance to the heat exchanger of the second circuit of the cogeneration system - 120°C and the temperature of the return water - 55°C, which supports the completion of the wood drying process using operation level 4 (Figure 4.5) and provides the possibility of entering the tolerance of operation level 1 regarding the change in the number of heat exchanger plates from 52 to 36 to support air heating for drying loaded fresh wood.

Chapter 5

Smart Grid Technology for Maintaining the Functioning of a Wind-Solar Electric System

Abstract

Integrated Smart Grid Systems of harmonization of production and consumption of electric power of the heat pump power supply and hot water power supply are developed. The integrated dynamic subsystem of the wind-solar electric system includes the following components: the electric network, wind turbine, photovoltaic solar panels, hybrid solar collectors, grid inverter, heat pump, two-section storage tank, upper section – for hot water supply, lower section – a low-grade energy source, frequency converter. Integrated systems based on predicting changes in the power factor, temperature of local water when measuring voltage from hybrid solar collectors at the input to the grid inverter, voltage at the output of the frequency converter and voltage frequency.

The adoption of advanced decisions to maintain the temperature of local water by changing the power of the electric motor of the heat pump compressors and electric motor of the circulation pump based on establishing the ratio of the voltage at the input to the grid inverter and the voltage at the output of the frequency converter are measured. The power factor of the wind-solar electric system is maintained. The complex mathematical and logical modeling of the wind-solar electric system, based on the mathematical substantiation of the architecture of the wind-solar electric system and mathematical substantiation of the maintenance of functioning of the wind-solar electric system, is performed.

Time constants and coefficients of the mathematical models of dynamics regarding the estimation of a change in the power factor of the system, the temperature of local water is predicted when measuring voltage from hybrid solar collectors at the input to the grid inverter, voltage at the output of the frequency converter and voltage frequency. Functional estimation of the change in power factor of the wind-solar electric system in the range of 58–98%. Functional estimation of a change in temperature of local water in the range of 30–55°C for hot water power supply and temperature of local water in the range of 35–55°C for heat pump power supply. Determining final functional information provides an opportunity to make forestalling decisions on a change in the of the electric motor of the heat pump compressors and electric motor of the circulation pump to prevent the peak load of the power system under voltage regulation conditions when connecting the heat pump power supply and hot water supply.

Keywords: wind-solar power system, photovoltaic solar panels, hybrid solar collectors

Introduction

The distributed generation of electricity using renewable sources requires intelligent systems for managing electricity flows and consumption. Smart Grid technoloies, demand management systems and energy storage are new components for the integration of distributed energy generation in the energy system. An urgent further development in this direction is predicting changes in the power factor, temperature of local water of heat pump power supply and hot water supply as part of a network wind-solar electric system using hybrid solar collectors.

In terms of connection to intelligent control systems, the author, Chaikovskaya, E. (2017) proposes to predict voltage changes when easuring the temperature of the electrolyte in the volume of batteries. An energy saving technology has been developed for the operation of a storage battery; it does not allow overcharging, but it does not allow discharge based on the coordination of electrochemical and diffusion processes of discharge and charge. The author's work (2019) is devoted to forecasting changes in parameters for connecting to Smart Grid technologies, which presents an integrated system for supporting the functioning of a wind-solar electric system based on predicting changes in battery capacity. The voltage is measured at the input to the hybrid charge controller and at the output from the inverter. The adoption of advanced decisions to change the power of the thermoelectric accumulator is based on establishing the ratio of the voltage at the input to the hybrid charge controller and at the output from the inverter when measuring the voltage frequency. The change in the flow rate and temperature of the heated water is provided, reducing the charge period to 30%, based on the change in the number of revolutions of the circulation pump electric motor.

So, in the work (Yinan, L., Wentao, Y., & Ping, H., Chang, Ch., Xiaonan, W., 2019) it is proposed to control the consumption of electrical energy by exchanging information in real time. The work (Saad, A., Samy, F., & Osama,

M., 2019) is devoted to the introduction of stochastic optimization of distributed generation of electrical energy using fuzzy logic. In (Perera, A., Vahid M., & Wickramasinghe, P., Scartezzini, J., 2019), a cyberphysical control system for the distributed generation of electrical energy, based on the theory of a consensus protocol, is proposed. In (Davye, M., Daranith, & Ch., Dae-Hyun, Ch., 2020), an intelligent converter for voltage regulation by absorbing or supplying reactive power is proposed. In (Xiqiao, L., Yukun, L., & Xianhong, B., 2019), a model for prioritizing sensitivity to changes in data is presented, based on accurate measurement of electrical energy consumption.

Methodological and Mathematical Substantiation

One of the main properties of energy systems is the mandatory exchange of substance, energy and information with the environment. Thus, the wind-solar electric system – open integrated dynamic system, the operation of which requires to predict changes in in the power factor, temperature of local water when measuring voltage from hybrid solar collectors at the input to the grid inverter, voltage at the output of the frequency converter and voltage frequency. The dynamic characteristics of the wind-solar electric system with sufficient accuracy for practice can be described by a finite set of parameters for changes in time, and the spatial coordinate that coincides with the direction of flow of the medium, such as changes power factor, temperature of local water. Therefore, the dynamic description of the the wind-solar electric system most fully and multifacetedly characterizes its operation. Thus, it is possible to determine that the real the wind-solar electric system – is a dynamic system, the mathematical model of which reflects the properties of the transformation of influences, i.e., its dynamic properties. Due to the fact that the the wind-solar electric system – reflects the dynamic peculiarities due to the nature of reactions to influences, the supporting of the functioning the wind-solar electric system – should be part of such a technological system, which is based on a dynamic system.

Based on the methodological, mathematical, logical substantiation of the technological systems (Chapter 1) the architecture, mathematical substantiation of the architecture (1), mathematical substantiation of maintenance of the operation (2) of the wind-solar electric system Smart Grid are proposed (Figure 5.1).

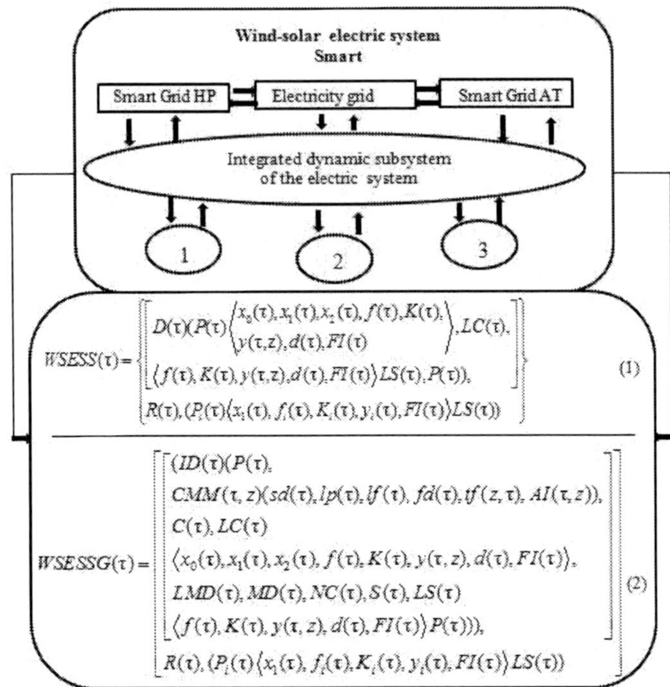

Figure 5.1. Wind-solar electric system Smart Grid: the architecture: HP – heat pump; AT – accumulator tank; 1 – charge unit; 2 – discharge unit; 3 – unit of evaluation of functional efficiency. Mathematical substantiation of the architecture (1). Mathematical substantiation of maintenance of the operation (2).

A wind-solar electric system is a dynamic system, the operation of which is the reproduction of a change in external, internal influences and initial conditions, for example, changes in solar radiation, wind speed, change in consumption of electrical energy and heat. Therefore, when designing a wind-solar electric system, an integrated dynamic subsystem is laid down in its base (Figure 5.1). The integrated dynamic subsystem includes the following components: mains, wind turbine, photovoltaic solar panels, hybrid solar collectors, grid inverter, heat pump, two-section storage tank, upper section – for hot water supply, lower section – a low-grade energy source, frequency converter. When representing the system design as the organization of a complex system, it was expanded by building up the dynamic subsystem blocks that forecast the process components around its base. Other components of the wind-solar electric system include the units of charge and discharge and functional efficiency estimation in a coordinated interaction with the dynamic subsystem (Figure 5.1). The mathematical substantiation of the architecture of the wind-solar electric system Smart (1), (Figure 5.1), based on the methodology of the mathematical description of dynamics of power systems, the method of the graph of cause-effect relations (Chapter 1) is proposed.

Where *WSESS* (τ) – wind-solar electric system Smart; τ – time, s; *ID*(τ) – integrated dynamic subsystem (mains, wind turbine, photovoltaic solar panels, hybrid solar collectors, grid inverter, heat pump, two-section storage tank, upper section – for hot water supply, lower section – a low-grade energy source, frequency converter); *P*(τ) – properties of the components of the wind-solar electric system; *x*(τ) – impacts (change in solar radiation, wind speed, change in consumption of electrical energy and heat; *f*(τ) – parameters that are measured: (voltage at the input to the grid inverter from hybrid solar collectors, voltage at the output of the frequency converter, frequency voltages); *K*(τ) – coefficients of mathematical description of the dynamics of power factor of the wind-solar electric system, the temperature of local water;; *y*(τ, *z*) – predicted output parameters (power factor, the temperature of local water) *z* – spatial coordinate of the condenser axis, heat exchanger coincides with the direction of the flow of motion of the medium, m; *d*(τ) – dynamic parameters of power factor change, the temperature of local water change; *FI*(τ) – functional resulting information on decision making; *LC*(τ) – logical relations regarding the control of the wind-solar electric system workability; *LS*(τ) – logical relations regarding the identification of the state of the wind-solar electric system; *R*(τ) – logical relations in *WSESS* (τ) to confirm the correctness of decisions made from the units of the wind-solar electric system. Indices: *i* – the number of elements of the wind-solar electric system; 0, 1, 2 – initial stationary mode, external, internal nature of impacts.

The mathematical substantiation of maintenance of the operation of the photoelectric charging station Smart Grid (2), (Figure 5.1), based on the methodology of the mathematical description of dynamics of power systems, the method of the graph of cause-effect relations (Chapter 1) is proposed. The basis of the proposed rationale is the mathematical description of the architecture of the wind-solar electric system Smart (1), (Figure 5.1). Prediction of changes in the power factor, temperature of local water when measuring voltage from hybrid solar collectors at the input to the grid inverter, voltage at the output of the frequency converter and voltage frequency enables making advanced decisions to maintain the temperature of local water by changing the power of the electric motor of the heat pump compressors and electric motor of the circulation pump based on establishing the ratio of the voltage at the input to the grid inverter and the voltage at the output of the frequency converter are measured. The power factor of the wind-solar electric system is maintained. The mathematical substantiation of Smart Grid maintenance of the operation of the wind-solar electric system (2) is proposed (Figure 5.1).

Where *WSESSG* (τ) – Smart Grid maintenance of the operation of the wind-solar electric system; τ – time, s; *ID*(τ) – integrated dynamic subsystem (mains, wind turbine, photovoltaic solar panels, hybrid solar collectors, grid inverter, heat pump, two-section storage tank, upper section – for hot water supply, lower section – a low-grade energy source, frequency converter). *P*(τ) – the properties of the elements of the integrated dynamic subsystem, units of the wind-solar electric system; *CMM*(τ, *z*) – complex mathematical modeling of the dynamics of changes in power factor, the temperature of local water; *sd*(τ) – the input data (the network wind-solar electric system with a capacity of 10 kW includes the following components: EUROWIND2 type (Ukraine) – wind power plant; ATMOSFERA-F2PV (Ukraine) – photovoltaic solar panels and hybrid solar panels. The heat pump system – Commotherm Hybrid Tower WW, Split DeLuxe (Austria) with a heating capacity of 5.7 kW is equipped with a two-section storage tank, the lower section of which has a volume of 200 liters, used as a low-potential energy source connected to hybrid solar collectors. The upper section of which has a volume of 300 liters used as a heat accumulator connected to hybrid solar collectors; *lp*(τ) – the boundary change in parameters (voltage from hybrid solar collectors at the input to the grid inverter, voltage at the output of the frequency converter and voltage frequency; *lf*(τ) – the levels of operation of the wind-solar electric system; *fd* (τ) – the obtained parameters (mode parameters of the wind-solar electric system); *tf*(τ,*z*) – the transfer function of predicted parameters – in the power factor, temperature of local water; *AI*(τ,*z*) – reference information for assessing changes in power factor, the temperature of local water; *C*(τ) – the control of workability of the wind-solar electric system; *LC*(τ) – the logical relations of the control of the wind-solar electric system workability; *x*(τ) – impacts (change in solar radiation, wind speed, change in consumption of electrical energy and heat; *f*(τ) – the measured parameters: (voltage at the input to the grid inverter from hybrid solar collectors, voltage at the output of the frequency converter, frequency voltages); *K*(τ) – coefficients of mathematical description of the dynamics of power factor of the wind-solar electric system, the temperature of local water;; *y*(τ, *z*) – predicted output parameters (power factor, the temperature of local water); *z* – spatial coordinate of the condenser axis, heat exchanger coincides with the direction of the flow of motion of the medium, m; $d(\tau)$ – dynamic parameters of power factor change, the temperature of local water change; $FI(\tau)$ – functional resulting information on decision making; $LMD(\tau)$ – the logical relations of decision making; $MD(\tau)$ – decision making; $NC(\tau)$ – the new conditions of the wind-solar electric system; $S(\tau)$ – the identification of the state of the wind-solar electric system; $LS(\tau)$ – the logical relations of identification of the state of the wind-solar electric system; $R(\tau)$ – the logical relations between the dynamic subsystem and units of charge, discharge, functional estimation of efficiency that belong to the wind-solar electric system. Indices: i – the number of elements of $WSESSG$ (τ); 0, 1, 2 – the initial, external, and internal character of influences.

Mathematical substantiation of the architecture of the the photoelectric charging station Smart (1) and mathematical substantiation of maintenance of the operation of the photoelectric charging station Smart Grid (2) (Figure 6.1) make it possible to maintain the operation of the wind-solar electric system using the following actions:

- Workability control ($C(\tau)$) of the dynamic subsystem based on complex mathematical ($CMM(\tau, z)$) and logical ($LC(\tau)$) modeling regarding obtaining standard ($AI(\tau,z)$) estimate of a change in the power factor of a change in the temperature of local water;
- Workability control ($C(\tau)$) of the dynamic system based on complex mathematical ($CMM(\tau, z)$) and logical ($LC(\tau)$) modeling regarding the obtaining functional ($FI\ (\tau)$) estimate of a change in the power factor of the wind-solar electric system of a change in the temperature of local water;
- Decision making ($MD(\tau)$) with the use of the functional resulting information ($FI\ (\tau)$), obtained based on logical modeling ($LMD(\tau)$)to maintain the temperature of local water by changing the power of the electric motor of the heat pump compressors and electric motor of the circulation pump based on establishing the ratio of the voltage at the input to the grid inverter and the voltage at the output of the frequency converter are measured;
- Identification ($S(\tau)$) of the new conditions of functioning of the the wind-solar electric system ($NC(\tau)$) based on logical modeling ($LS(\tau)$) as a part of the dynamic subsystem and confirmation of new operating conditions based on logical modeling ($R(\tau)$) from the units of the the wind-solar electric system.

Complex Mathematical Modeling of Heat Pump Power Supply and Hot Water Supply Using Hybrid Solar Collectors

According to Formulas (1), (2), the prediction of a change in the power factor of the wind-solar electric system and the temperature of local water was proposed. The voltage at the inlet to the grid inverter from hybrid solar collectors, and at the outlet from the frequency converter and voltage frequency is measured. The transfer function by the channel "power factor of the wind-solar electric system – voltage at the inlet to the grid inverter" is complex. A change in power factor and the local water temperature are estimated. A change in the local water temperature is estimated both over time and along the spatial coordinate of the condenser axis, heat exchanger coincides with the direction of the flow of motion of the medium. The transfer function along the channel "power factor of the wind-solar electric system – voltage at the inlet to the grid inverter," which was obtained as a result of solving a system of nonlinear differential equations, is presented as follows:

$$W_{PF-U_1} = \frac{K_{pf} K_t \varepsilon \left(1 - L_e^*\right)}{\left(T_w S + 1\right)\beta - 1}\left(1 - e^{-\gamma\xi}\right), \tag{3}$$

where

$$K_{pf} = \frac{I(U_1 - U_2)}{(N_e)}; \; K_t = \frac{m(\theta_0 - \sigma_0)}{G_{i0}}; \; \varepsilon = \frac{\alpha_{e0} h_{e0}}{\alpha_{i0} h_{i0}};$$

$$L_e^* = \frac{1}{L_e + 1}; \; L_e = \frac{G_e C_e}{\alpha_{e0} h_{e0}}; \; T_w = \frac{g_w C_w}{\alpha_{i0} h_{i0}}; \; \beta = T_m S + \varepsilon^* + 1;$$

$$\varepsilon^* = \varepsilon(1 - L_e^*); \; \gamma = \frac{(T_w S + 1)\beta - 1}{\beta}; \; \xi = \frac{z}{L_i}; \; L_i = \frac{G_i C_i}{\alpha_{i0} h_{i0}}.$$

where PF is the power factor of the wind-solar electric system; I – current, A; U_1, U_2 – voltage at the input to the grid inverter and at the output from the frequency converter, respectively, V; Ne – the power of the wind-solar electric system, kW; C is the specific thermal capacity, kJ/(kg·K); α is the heat transfer factor, kW/(m^2·K); G is the loss of substance, kg/s; g is the specific weight of a substance, kg/m; h is the specific surface, m^2/m; σ, θ are the temperature of the warming heat carrier and of the separating wall, respectively, K; z is spatial coordinate of the condenser axis, heat exchanger coincides with the direction of the flow of motion of the medium m; T_w, T_m are the time constants that characterize the thermal accumulating of local water, metal, s; m is the indicator of the dependence of heat transfer factor on consumption; τ is the time, s; S is the Laplace parameter; $S = \omega j$; ω is the frequency, 1/s. Indices: 0 – initial stationary mode; i – internal flow – local water; e – external flow – warming heat carrier; m – metal wall.

A real part of the transfer function was separated:

$$O(\omega) = \frac{(L_1 A_1) + (M_1 B_1) K_{pf} K_t \varepsilon (1 - L_e^*)}{(A_1^2 + B_1^2)}. \tag{4}$$

The K_t factor includes the temperature of the separating wall θ:

$$\theta = \left(\alpha_i \frac{(\sigma_1 + \sigma_2)}{2} + A \frac{(t_1 + t_2)}{2} \right) / (\alpha_i + A), \tag{5}$$

where σ_1, σ_2 are the temperatures of warming heat carrier at the inlet and at the outlet of the heat exchanger, K, respectively; t_1, t_2 are the temperatures of local water at the inlet and at the outlet of the heat exchanger, K, respectively; α is the heat transfer factor, $kW/(m^2 \cdot K)$. Index: i – local water.

$$A = \frac{1}{(\delta_m / \lambda_m + 1/\alpha_e)}, \tag{6}$$

where δ is the thickness of a wall of the heat exchanger, m; λ is the thermal conductivity of the metal wall of the heat exchanger, kW/(m·K). Indices: e – warming heat carrier; m – metal wall of a heat exchanger.

To use the real part $O(\omega)$, the following factors were obtained:

$$A_1 = \varepsilon^* - T_w T_m \omega^2; \tag{7}$$

$$A_2 = \varepsilon^* + 1; \tag{8}$$

$$B_1 = T_w \varepsilon\omega + T_w\omega + T_m\omega; \tag{9}$$

$$B_2 = T_m\omega; \tag{10}$$

$$C_1 = \frac{A_1 A_2 + B_1 B_2}{A_2^{\ 2} + B_2^{\ 2}}; \tag{11}$$

$$D_1 = \frac{A_2 B_1 - A_1 B_2}{A_2^{\ 2} + B_2^{\ 2}}; \tag{12}$$

$$L_1 = 1 - e^{-\zeta C_1} \cos(-\xi D_1); \tag{13}$$

$$M_1 = -e^{-\zeta C_1} \sin(-\xi D_1). \tag{14}$$

The transfer function (3), which was obtained based on the use of the operator method of solving the system of nonlinear differential equations, retains the Laplace transform parameter – S ($S = \omega j$), where ω is the frequency, 1/s. To switch from the frequency area to the time area, a real part (4), obtained as a result of the mathematical treatment of transfer function, was separated. It is this part that is included in the integrals (15), (16), which makes it possible to obtain dynamic characteristics of a change in power factor of the wind-solar electric system, the temperature of local water using the inverse Fourier transform:

$$PF(\tau) = \frac{1}{2\pi} \int_0^\infty K_{\text{pf}} K_{\text{t}} O(\omega) \sin(\tau\omega/\omega) d\omega, \tag{15}$$

$$t(\tau, z) = \frac{1}{2\pi} \int_0^\infty K_{\text{pf}} K_{\text{t}} O(\omega) \sin(\tau\omega/\omega) d\omega, \tag{16}$$

where PF is the power factor of the wind-solar electric system; t is the temperature of local water, K.

So, for obtaining the reference estimation of change in the power factor of the wind-solar electric system; the temperature of local water a block diagram is proposed (Figure 5.2) using, for example, the initial data of a grid-type wind-solar electric system with a power of 10 kW using hybrid solar collectors to support the functioning of a heat pump power supply and hot water supply.

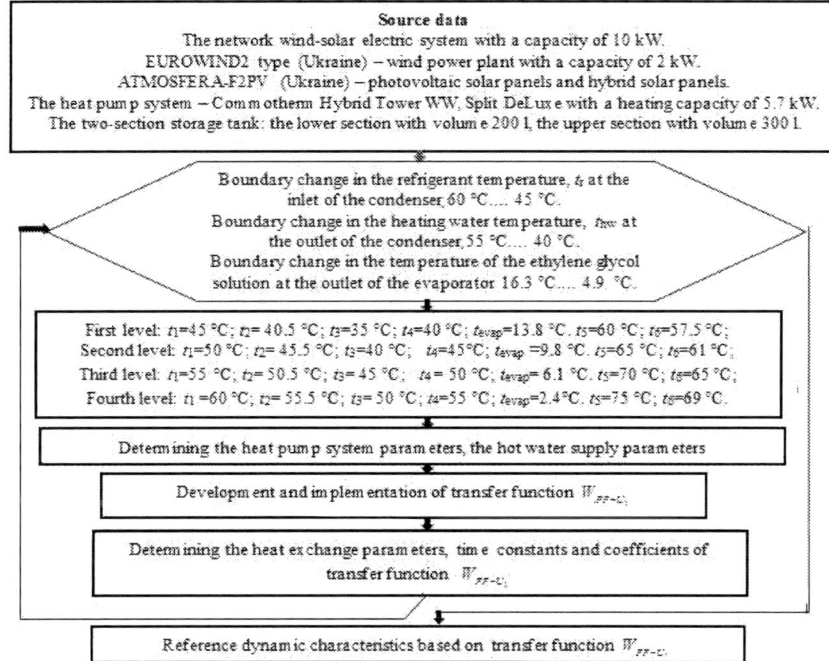

Figure 5.2. Block diagram of comprehensive mathematical modeling of the wind-solar electric system: t_1, t_2 – the refrigerant temperature at the inlet of the condenser and the outlet of the condenser, respectively, °C; t_3, t_4 – the local water temperature at the inlet of the condenser and the outlet of the condenser, respectively, °C; t_5, t_6 – the temperature of the water from hybrid solar collectors at the inlet to the heat accumulator and at the outlet from the heat accumulator, respectively, °C; t_{evap} – the refrigerant evaporation temperature.

The network wind-solar electric system with a capacity of 10 kW includes the following components: EUROWIND2 type (Ukraine) – wind power plant; ATMOSFERA-F2PV (Ukraine) – photovoltaic solar panels and hybrid solar panels. The heat pump system – Commotherm Hybrid Tower WW, Split DeLuxe (Austria) with a heating capacity of 5.7 kW is equipped with a two-section storage tank, the lower section of which has a volume of 200 liters, used as a low-potential energy source connected to hybrid solar collectors. The upper section of which has a volume of 300 liters used as a heat accumulator connected to hybrid solar collectors. The following levels of operation of the heat pump system have been established for the change in the temperature of the refrigerant at the inlet to the condenser and at the outlet from the condenser: – first level: 45–40.5°C; second level: 50–45.5°C; third level: 55–50.5°C; fourth level: 60–55.5°C. They correspond to changes in the temperature of local water: 35–40°C; 40–45°C; 45–50°C; 50–55°C. The temperature of the low-potential energy source – ethylene glycol solution at the outlet of the heat pump evaporator is: 16.3°C, 12.3°C, 8.6°C, 4.9°C, corresponding to the established operating levels for heating up to a temperature of 20°C. The following levels of operation of the hot water supply have been established for the change in the temperature of the water from hybrid solar collectors at the inlet to the heat accumulator and at the outlet from the heat accumulator: first level: 60–57.5°C; second level: 65–61°C; third level: 70–65°C; fourth level: 75–69°C. They correspond to changes in the temperature of local water: 30–55°C. Warming heat carrier consumption – 0.095 kg/s. One of the most problematic areas is the harmonization of energy production and consumption in the context of distributed energy generation using renewable sources. Connecting to Smart Grid technologies will prevent the peak load of the power system in conditions of voltage regulation when connecting the heat pump power supply and hot water power supply.

Table 1. Operating parameters of the heat pump system

Levels of operation	G_r, kg/s	N_e, kW	N_t, kW	U, V	f, Hz	n, rpm	COP
First level	0.0340	0.705	1.4	194.2	24.28	738.4	8.08
Second level	0.0350	0.923	2.9	254.3	31.78	953.4	6.17
Third level	0.0357	1.185	4.3	326.5	40.81	1224.3	4.80
Fourth level	0.0363	1.452	5.7	400	50	1500	3.92

Note: G_r – refrigerant consumption, kg/s; N_e – power of the compressor electric motor, kW; N_t – thermal power of the low-grade energy source, kW; U – voltage, V; f – voltage frequency, Hz; n – the number of revolutions of the compressor electric motor, rpm; COP – coefficient of performance of the heat pump system.

Table 2. Operating parameters of the hot water supply

Levels of operation	G_w, kg/s	N_t, kW	U, V	f, Hz	n, rpm
First level	0.0090	1	92	20	1140
Second level	0.0100	1.5	138	30	1710
Third level	0.0106	2	184	40	2280
Fourth level	0.0120	2.5	230	50	2850

Note: Gw – local water consumption, kg/s; Nt – heating power, kW; U – voltage, V; f – voltage frequency, Hz; n – the number of revolutions of the circulation pump electric motor, rpm.

According to formulas (1) - (3), the results of complex mathematical modeling of the heat pump power supply and hot water supply using hybrid solar collectors are presented (Tables 1, 2, 3 - 6).

Table 3. Heat transfer parameters as part of complex mathematical modeling of heat pump power supply

Levels of operation	Parameter		
	α_r, kW/(m²·K)	α_w, kW/(m²·K)	k, kW/(m²·K)
First level	1.15	0.785	0.460
Second level	1.19	0.901	0.505
Third level	1.26	1.076	0.570
Fourth level	1.38	1.372	0.674

Note: α_r – coefficient of convective heat transfer from the refrigerant to the condenser wall, kW/(m²·K); α_w – coefficient of convective heat transfer from the condenser wall to local water, kW/(m²·K); k – heat transfer coefficient, kW/(m²·K).

Table 4. Heat transfer parameters as part of complex mathematical modeling of the hot water supply

Levels of operation	Parameter		
	α_c, kW/(m²·K)	α_w, kW/(m²·K)	k, kW/(m²·K)
First level	1.513	0.749	0.486
Second level	1.533	0.851	0.530
Third level	1.549	0.927	0.560
Fourth level	1.555	0.953	0.564

Note: α_c – coefficient of convective heat transfer from the warming heat carrier to the heat exchanger wall, kW/(m²·K); α_w – coefficient of convective heat transfer from the heat exchanger wall to local water, kW/(m²·K); k – heat transfer coefficient, kW/(m²·K).

Table 5. Time constants and coefficients of the mathematical model of the dynamics of the temperature of the water of the heat pump power supply

Levels of operation	T_w, s	T_m, s	L_w, m	ε	L_r, m	L_r^*	ε^*	ζ
First level	6.23	2.62	24.61	1.7756	2.50	0.2857	1.2683	0.6811
Second level	5.42	2.28	21.43	1.6014	2.48	0.2874	1.1412	0.6309
Third level	4.54	1.91	17.95	1.4182	2.44	0.2907	1.0059	0.5945
Fourth level	3.56	1.50	14.08	1.2204	2.30	0.3030	0.8506	0.5794

Note: T_w, T_m – time constants characterizing the thermal storage capacity of local water, metal, s; L_w, L_r, L_r*, ε, ε*, ζ – coefficients of the mathematical model of the local water temperature dynamics.

Table 6. Time constants and coefficients of the mathematical model of the dynamics of the temperature of the water of the hot water power supply

Levels of operation	T_w, s	T_m, s	L_w, m	ε	L_c, м	L_c^*	ε*	ζ
First level	37.4	7.96	0.59	1.72	3.65	0.22	1.34	2.8
Second level	32.9	7.0	0.58	1.53	3.61	0.22	1.19	2.85
Third level	30.2	6.43	0.56	1.42	3.57	0.22	1.11	2.95
Fourth level	29.4	6.26	0.62	1.39	3.55	0.22	1.08	2.66

Note: T_w, T_m – time constants characterizing the thermal storage capacity of local water, metal, s; L_w, L_c, L_c*, ε, ε*, ζ – coefficients of the mathematical model of the local water temperature dynamics.

As presented in Tables 5, 6, the time constants and coefficients that are part of the mathematical model of dynamics (3) are obtained on the basis of parameters as part of complex mathematical modeling (Tables 1–4).

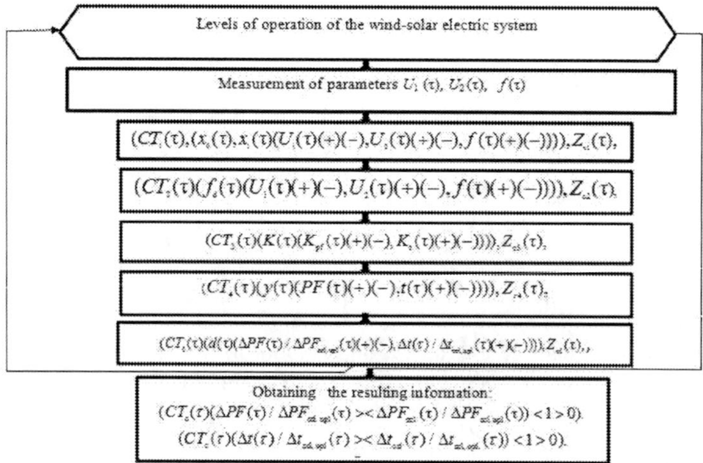

Figure 5.3. Block diagram of the wind-solar electric system functioning control: U_1, U_2 – voltage at the input to the network inverter and at the output from the frequency converter, V; f – voltage frequency, Hz; PF – power factor of the wind-solar electric system; t – local water temperature, °C; CT – event control; Z – logical relations; d – dynamic parameters; x – effects; f_d – parameters measured; y – parameters predicted; K – coefficients of mathematical description; ι – time, s. Indices: c – control of operability; ccl upl – constant calculated value of the parameter of lower level, upper level of operation (heat pump power supply), (hot water power supply), respectively; ccl – constant calculated value of the level of operation parameter; 0, 1, 2 – initial stationary mode, external, internal influences; 3 – coefficients of dynamics equations; 4 – significant predicted parameters; 5 – dynamic parameters.

Based on the proposed mathematical substantiation Smart Grid maintenance of functioning of the wind-solar electric system (1) to (4) the block diagram for the control of workability of the wind-solar electric system (Figure 5.3) is developed.

Control of workability of the wind-solar electric system (Figure 5.3) enables obtaining the resulting information on decision-making about the maintenance of operation of the wind-solar electric system.

Maintaining of Voltage in the Distribution System Based on Coordination of Production and Consumption of Energy

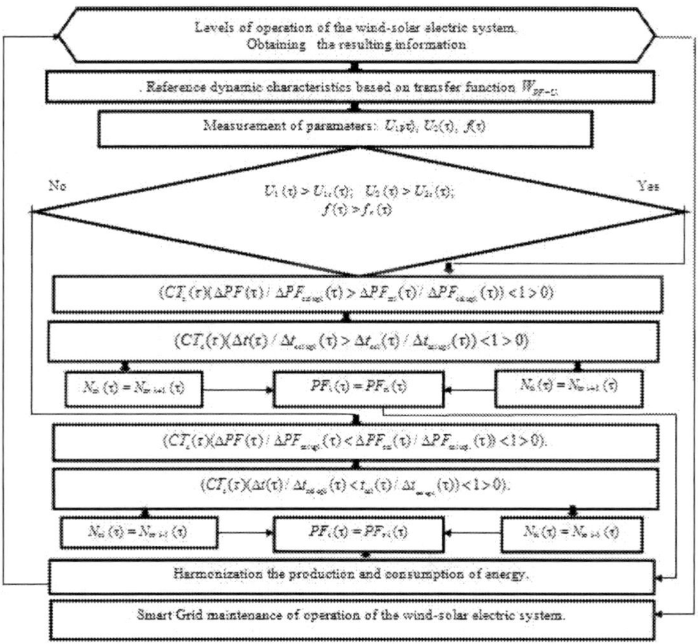

Figure 5.4. Block diagram of maintenance of operation of the wind-solar electric system: U_1, U_2 – voltage at the input to the network inverter and at the output from the frequency converter, V; f – voltage frequency, Hz; PF – power factor of the wind-solar electric system; N_e, N_t – power of the compressor electric motor, heating power, respectively, kW; ι – time, s; Indices – i – number of operation levels; r – reference value of the parameter; ccl upl – constant calculated value of the parameter of lower level, upper level of operation (heat pump power supply), (hot water power supply), respectively; ccl – constant calculated value of the level of operation parameter.

Table 5.7. Integrated Smart Grid System of harmonization of production and consumption of electric power of the heat pump power supply

Time, τ, 10^3 s	Changing the parameters of the technological process	$\Delta PF(\tau)/\Delta PF_1(\tau)$	$PF(\tau)$	$t_w(\tau)$, °C
0	Charge $U_1 = 84$ V; $U_2 = 190.7$ V; $f = 25.09$ Hz; $G_{r.} = 0.0340$ kg/s; $t_{r\ in} = 45$°C; $t_{r\ out.} = 40.5$°C; $t_{w\ in} = 35$°C; $t_{w\ out.} = 40$°C	1	0.58	35
10.5	$U_1 = 90$ V $U_2 = 190.7$ V; $t_{r\ in} = 45$°C; $t_{r\ out} = 40.5$°C; $t_{w\ in} = 35$°C; $t_{w\ out.} = 40$°C	0.9438	0.6025	36.12
21	$U_1 = 96$ V; $U_2 = 190.7$ V $t_{r\ in} = 45$°C; $t_{r\ out} = 40.5$°C; $t_{w\ in} = 35$°C; $t_{w\ out.} = 40$°C	0.8875	0.6250	37.25
31.5	$U_1 = 108$ V; $U_2 = 190.7$ V $t_{r\ in} = 45$°C; $t_{r\ out} = 40.5$°C; $t_{w\ in} = 35$°C; $t_{w\ out.} = 40$°C	0.7751	0.6700	39.5
42	$U_1 = 114$ V; $U_2 = 190.7$ V $t_{r\ in} = 45$°C; $t_{r\ out} = 40.5$°C; $t_{w\ in} = 35$°C; $t_{w\ out.} = 40$°C	0.7188	0.6925	40.63
52.5	$U_1 = 126$ V; $U_2 = 190.7$ V $t_{r\ in} = 45$°C; $t_{r\ out} = 40.5$°C; $t_{w\ in} = 35$°C; $t_{w\ out.} = 40$°C	0.6064	0.7375	42.88
63	$U_1 = 130$ V; $U_2 = 190.7$ V $t_{r\ in} = 45$°C; $t_{r\ out} = 40.5$°C; $t_{w\ in} = 35$°C; $t_{w\ out.} = 40$°C	0.5689	0.7525	43.63
73.5	$U_1 = 136$ V; $U_2 = 190.7$ V $t_{r\ in} = 45$°C; $t_{r\ out} = 40.5$°C; $t_{w\ in} = 35$°C;$t_{w\ out.} = 40$°C	0.5126	0.7750	44.76
84	$U_1 = 140$ V; $U_2 = 190,7$ V $t_{r\ in} = 45$°C; $t_{r\ out} = 40.5$°C; $t_{w\ in} = 35$°C; $t_{w\ out.} = 40$°C	0.4752	0.7900	45.51
94.5	Decision making on the discharge $G_r = 0.035$ kg/s; $U_1 = 173.96$ V; $U_2 = 241.56$ V; $f = 31.78$ Hz; $t_{r\ in} = 50$°C; $t_{r\ out.} = 45.5$°C; $t_{w\ in} = 40$°C; $t_{w\ out.} = 45$°C	0.4783	0.7912	45.51

Time, τ, 10^3 s	Changing the parameters of the technological process	$\Delta PF(\tau)/\Delta PF_1(\tau)$	$PF(\tau)$	$t_w(\tau)$, °C
105	Charge $U_1 = 179.96$ V; $U_2 = 241.56$ V; $f = 31.78$ Hz; $t_{r\,in} = 50$°C; $t_{r\,out} = 45.5$°C; $t_{w\,in} = 40$°C; $t_{w\,out} = 45$°C	0.4355	0.8083	46.37
115.5	$U_1 = 185.96$ V; $U_2 = 241.56$ V; $t_{r\,in} = 50$°C; $t_{r\,out} = 45,5$°C; $t_{w\,in} = 40$°C; $t_{w\,out} = 45$°C	0.3931	0.8253	47.22
126	$U_1 = 191.96$ V; $U_2 = 241.56$ V $t_{r\,in} = 50$°C; $t_{r\,out} = 45.5$°C; $t_{w\,in} = 40$°C; $t_{w\,out} = 45$°C	0.3506	0.84	48.07
136.5	$U_1 = 197.96$ V; $U_2 = 241.56$ V $t_{r\,in} = 50$°C; $t_{r\,out} = 45,5$°C; $t_{w\,in} = 40$°C; $t_{w\,out} = 45$°C	0.3082	0.8570	48.92
147	Decision making on the discharge $G_r = 0.0357$ kg/s; $U_1 = 257.9$ V; $U_2 = 310.17$ V; $f = 40.81$ Hz; $t_{r\,in} = 55$°C; $t_{r\,out} = 50.5$°C; $t_{w\,in} = 45$°C; $t_{w\,out} = 50$°C	0.3125	0.8586	49
157.5	Charge $U_1 = 281.9$ V; $U_2 = 310.17$ V; $f = 40.81$ Hz; $t_{r\,in} = 55$°C; $t_{r\,out} = 50.5$°C; $t_{w\,in} = 45$°C; $t_{w\,out} = 50$°C	0.1690	0.9154	51.87
161.6	Decision making on the discharge $G_r = 0.0363$ kg/s; $U_1 = 341.93$ V; $U_2 = 380$ V; $f = 50$ Hz; $t_{r\,in} = 60$°C; $t_{r\,out} = 55.5$°C; $t_{w\,in} = 50$°C; $t_{w\,out} = 55$°C	0.0690	0.9554	53.67
161.6	Discharge $U_1 = 341.93$ V; $U_2 = 380$ V; $f = 50$ Hz; $G_r = 0.0363$ kg/s; $t_{r\,in} = 60$°C; $t_{r\,out} = 55.5$°C; $t_{w\,in} = 50$°C; $t_{w\,out} = 55$°C	0	0.98	55

Note: PF – power factor; $t_{r\,in}$, $t_{r\,out}$, $t_{w\,in}$, $t_{w\,out}$ – refrigerant temperature, local water temperature at the inlet to the condenser, at the outlet from the condenser, °C; G_r – refrigerant consumption, kg/s; f – voltage frequency, Hz; U_1, U_2 – voltage at the input to the network inverter and at the output from the frequency converter, V; τ – time, s. Indexes: w – internal flow – local water; 1 – constant, calculated value of the parameter of the lower level of functioning.

Based on the proposed mathematical substantiation Smart Grid (1) to (4) (Figure 5.3) the block diagram of maintenance of operation of the wind-solar electric system (Figure 5.4) is developed.

The coordination of production and consumption of energy (Figure 5.4) makes it possible to ensure the operation of the wind-solar electric system.

Results and Discussion

Harmonization of Production and Consumption of Electric Power of the Heat Pump Power Supply

A complex integrated system for supporting the functioning of heat pump power supply has been developed (Table 5.7). This system is based on the predicted power factor and local water temperature changes. The voltage at the input to the grid inverter, the voltage at the output from the frequency converter and the voltage frequency are continuously measured. Advance decisions are made to change the power of the heat pump compressor electric motor in accordance with the change in the thermal power of the lower section of the two-section storage tank as a low potential energy source.

A power factor of the wind-solar electric system in the established period is determined as follows:

$$PF_{i+1}(\tau) = PF_i + \\ + \begin{pmatrix} \Delta PF_{i+1}(\tau) / \Delta PF_{1..}(\tau) - \\ -\Delta PF_i(\tau) / \Delta PF_1(\tau) \end{pmatrix} (PF_2 - PF_1), \tag{17}$$

where PF is the power factor of the wind-solar electric system; PF_1, PF_2 are the initial, final values of power factor; τ is the time, s. Index: 1 is the constant, calculated value of the parameter of the lower level of functioning; i is the number of levels of the wind-solar electric system operation.

The temperature of local water in the established period is determined as follows:

$$t_{wi+1}(\tau) = t_{wi} + \left(\left(\frac{\Delta t_{wi+1}(\tau)}{\Delta t_{1.}(\tau)} - \frac{\Delta t_{wi}(\tau)}{\Delta t_1(\tau)} \right) (t_{w2} - t_{w1}) \right), \tag{18}$$

where t_w – temperature of local water, °C; t_1, t_2 – initial and final value of local water temperature, °C, respectively; i – the number of levels of functioning; τ – time, s. Index 1 – constant, calculated value of the parameter of the lower level of functioning.

So, for example, in the period of time $147 \cdot 10^5$ s (4083 h) from the beginning of the heating season, the predicted increase in the power factor went from 0.8570 to 0.8586. The power factor using formula (17) is:

$$0.8586 = 0.8570 + (0.3125\text{-}0.3080)(0.98\text{-}0.95)$$

In the period of time $147 \cdot 10^5$ s (4083 h), the predicted increase in the temperature of local water went from 48.92°C to 49°C. The temperature of local water using formula (18) is:

$$49°C = 48.92°C + (0.3125 - 0.3080)(55°C - 35°C).$$

During this period of time, when the voltage at the input to the grid inverter changes to a level of 257.9 V, it is necessary to make a decision to maintain the power factor of the wind-solar electrical system at the level 0.8586.

Predicting an increase in the electric grid charge, it is necessary to increase the power of the heat pump compressor electric motor based on the voltage frequency change to 40.81 Hz. Adopting a proactive decision to increase the number of revolutions of the compressor electric motor allows increasing the refrigerant consumption to the level of 0.0357 kg/s and ensuring the maintenance of the local water temperature at 49°C. The implementation of such actions will allow, while maintaining the functioning of the heat pump power supply, to coordinate the production and consumption of energy.

Harmonization of Production and Consumption of Electric Power of the Hot Water Power Supply

A complex integrated system for supporting the functioning of the hot water power supply has been developed (Table 5.8).

Table 5.8. Integrated Smart Grid System of harmonization of production and consumption of electric power of the hot water power supply

Time, τ, 10^3 s	Changing the parameters of the technological process	$\Delta PF(\tau)/\Delta PF_1(\tau)$	$PF(\tau)$	t_w (τ), °C
3	Charge $U_1 = 32$ V; $U_2 = 38.8$ V; $f = 8.43$ Hz; $G_w = 0.009$ kg/s; $t_{in} = 60$°C; $t_{out} = 57.5$°C	0.1435	0.6374	33.59
6	$U_1 = 72$ V; $U_2 = 86.27$ V; $f = 18.75$ Hz; $G_w = 0.009$ kg/s; $t_{in} = 60$°C; $t_{out} = 57.5$°C	0.3010	0.7004	37.53
9	Decision making on the discharge $G_w = 0.0100$ kg/s;$U_1 = 72$ V; $U_2 = 86.27$ V; $f = 18.75$ Hz;$t_{in} = 65$°C; $t_{out} = 61$°C	0.3256	0.7102	38.14
12	Charge $U_1 = 108$ V; $U_2 = 129.4$ V; $f = 28.13$ Hz; $G_w = 0.010$ kg/s;$t_{in} = 65$°C; $t_{out} = 61$°C	0.4883	0.7753	42.21
15	Decision making on the discharge $G_w = 0.0106$ kg/s;$U_1 = 108$ V; $U_2 = 129.4$ V; $f = 28.13$ Hz;$t_{in} = 70$°C; $t_{out} = 65$°C	0.5567	0.8027	43.92
18	Charge $U_1 = 144$ V; $U_2 = 172.5$ V; $f = 37.5$ Hz; $G_w = 0.0106$ kg/s;$t_{in} = 70$°C; $t_{out} = 65$°C	0.7415	0.8766	48.54
21	$U_1 = 180$ V; $U_2 = 215.7$ V; $f = 46.9$ Hz; $G_w = 0.0106$ kg/s;$t_{in} = 70$°C; $t_{out} = 65$°C	0.9288	0.9515	53.22
24	Decision making on the discharge $G_w = 0.0120$ kg/s;$U_1 = 180$ V; $U_2 = 215.7$ V; $f = 46.9$ Hz;$t_{in} = 75$°C; $t_{out} = 69$°C	0.9395	0.9558	53.49
27	Discharge $U_1 = 182$ B; $U_2 = 230$ B; $f = 50$ Hz; $G_w = 0.0120$ kg/s; $t_{in} = 75$°C; $t_{out} = 69$°C	1	0.98	55

Note: *PF* – power factor; tin, tout– warming heat carrier temperature at the inlet to the the heat exchanger, at the outlet from the the heat exchanger, °C; t_w – local water temperature, °C; G_w – local water consumption, kg/s; f – voltage frequency, Hz; U_1, U_2 – voltage at the input to the network inverter and at the output from the frequency converter, V; τ – time, s. Indexes: w – local water; 1 – constant, calculated value of the parameter of the upper level of functioning.

This system is based on the predicted power factor and local water temperature changes. The voltage at the input to the grid inverter, the voltage at the output from the frequency converter and the voltage frequency are continuously measured. Advance decisions are made to change the power of the electric motor of the circulation pump, based on the change in the voltage frequency in accordance with the change in the thermal power of the upper section of the two-section storage tank.

So, for example, in the period of time $15 \cdot 10^3$ s (4.2 h) from the beginning of the solar radiation's predicted increase power factor from 0.7753 to 0.8027. The power factor using formula (17) is:

$$0.8027 = 0.7753 + (0.5567 - 0.4883)(0.98 - 0.95).$$

In this period of time $15 \cdot 10^3$ s (4.2 h), the predicted increase in the temperature of local water went from 42.21°C to 43.92°C. The temperature of local water using formula (18) is:

$$43.92°C = 42.21°C + (0.5567 - 0.4883)(55°C – 30°C).$$

During this period of time, when the voltage at the input to the grid inverter changes to a level of 108 V, it is necessary to make a decision to maintain the power factor of the wind-solar electrical system at the level 0.8027. Predicting an increase in the electric grid charge, it is necessary to increase the power of the circulating pump electric motor based on the voltage frequency change to 28.13 Hz. Adopting a proactive decision to increase the number of revolutions of the electric motor allows increasing the local water consumption to the level of 0.0106 kg/s and ensuring the maintenance of the local water temperature at 43.92°C. The implementation of such actions will allow, while maintaining the functioning of the hot water power supply, to coordinate the production and consumption of energy.

Chapter 6

Smart Grid Technology for Maintaining the Functioning of Photoelectric Charging Stations

Abstract

Integrated Smart Grid Systems of harmonization of production and consumption of electric power based on a prediction of changes in the battery capacity is developed. The integrated dynamic subsystem the photoelectric charging station includes the following components: mains, photoelectric solar panels, a hybrid inverter, rechargeable batteries, a two-way Smart Meter counter and a charger. Advanced decisions on the change in power transmission capacity have made it possible to regulate voltage in the distribution system by maintaining the power factor of the photoelectric charging station. Voltages at the input to the hybrid inverter and in the distribution system were measured to assess their ratio. Comprehensive mathematical and logical modeling of the photoelectric charging station was performed based on the mathematical substantiation of architecture and operation maintenance.

A dynamic subsystem including such components as mains, a photoelectric module, a hybrid inverter, batteries, a two-way counter Smart Meter and a charger formed the basis of the proposed technological system. Time constants and coefficients of mathematical models of dynamics in terms of estimation of changes in the battery capacity and power factor of the photoelectric charging station were determined. A functional estimate of changes in the battery capacity and power factor of the photoelectric charging station was obtained. Maintenance of voltage in the distribution system was realized based on resulting operation data to estimate a change in the battery capacity. Advanced decision-making has made it possible to raise the power factor of the photoelectric charging station up to 40% due to matching the electric power production and consumption. Maintenance of operation of the photoelectric charging station using the developed Smart Grid technology has enabled prevention of peak loading of the power system due to a 20% reduction of power consumption from the network.

Keywords: photoelectric charging station, rechargeable battery, hybrid inverter, two-way counter Smart Meter

Introduction

A serious complication occurring in the use of AC charging stations consists of a risk of peak loading of mains. In conditions of the growing number of electric vehicles and irregularity of charge, there is a need to build charging stations using renewable energy sources. Distributed generation of electric energy using renewable sources requires connection to Smart Grid technologies for integration into the electric network. Maintaining of power factor of the photoelectric charging station as regards redistribution of produced and consumed electricity is an urgent problem of further development of Smart Grid technologies. To this end, it is necessary to predict changes in battery capacity and power factor of the photoelectric charging station when measuring the input voltage of a hybrid inverter and voltage in the distribution system to assess their ratio.

Advanced decisions concerning change in the level of power transmission to the electric network make it possible to adjust the voltage in the distribution system to maintain the balance of active and reactive power without the use of additional devices. The power factor of photoelectric charging stations is maintained by coordination of production and consumption of electric energy. This enables the prevention of peak loads in the electric network in conditions of satisfaction of growing consumer demands. For example, in author's work (E. Chaikovskaya, 2017) based on predicting voltage changes in a storage battery is devoted to connection to Smart Grid technologies. A technology of maintaining change in the capacity of a storage battery during the measurement of electrolyte temperature in a set of accumulators was presented. The use of an integrated system of estimating a change in voltage based on matching electrochemical and diffusion processes of charge and discharge enables making advanced decisions on boosting to prevent impermissible overcharge and discharge.

The author's work (E. Chaikovskaya, 2019) tackles forecasting change in the battery capacity for connection to Smart Grid technologies. It presents an integrated system of maintaining the operation of a wind-solar electrical system. Making advanced decisions on changing the power of a thermoelectric accumulator is based on establishing a ratio between the voltage measured at the input to a hybrid charge controller and at an inverter output when measuring frequency. Up to 30% reduction of thermoelectric accumulator charge time was provided based on a change of the number of turns of a circulating pump motor to change flow rate and temperature of heated water.

Optimization of a charging station integrated into a distribution network with the use of renewable energy sources was presented in (Ferro, G., Laureri, F., Minciardi, R., Robba, M., 2019). A target function has been developed. Its minimization is based on the sum of the costs of charging electric vehicles from an external network and the costs associated with service delays. The need to develop infrastructure in a connection with establishing the optimal location of charging stations was stated in (Jordán, J., Palanca, J., del Val, E., Julian, V., Botti, V., 2021). An agent-oriented approach based on a genetic algorithm was presented. The proposed multi-agent system takes into account the data of activities in social networks and information on mobility to establish optimal configurations.

A strategy of hybrid power production and charging electric cars was developed in (Jiao, Z., Lu, M., Ran, L., Shen, Z.-J. M., 2020) on the basis of an integrated stochastic model of planning. Queue optimization and planning volume of power generated from non-renewable and renewable energy sources taking into account the non-stability of solar radiation was used. A model of multipurpose optimization of a charging station based on the theory of fuzzy numbers was proposed in (Liu, J., Dai, Q., 2020). An algorithm of a swarm of particles was presented to determine the optimal operation of charging stations with a possibility of testing the model in real operating conditions. Optimization of the charging station operation based on mixed integer programming was presented in (Zaher, G. K., Shaaban, M. F., Mokhtar, M., Zeineldin, H. H. 2021). It was solved in a form of a diagram for the day ahead.

The purpose of the proposed approach implies maximizing the profit of the charging station owner while satisfying power consumers based on data of charge and discharge and battery replacement during the day. This model takes into account the arrival of customers, changes in the price of electricity from the net, restrictions on connection to the net, and self-destruction of batteries. The work (Elma, O., 2020) addresses the improvement of power consumption management. A technology of fast charge with direct current based on dynamic estimation of a hybrid station was offered. A decrease in a peak load on the power system during periods of electric vehicle charge and an increase in battery life due to more controlled coordination of the discharge/charge have been established. It was proposed in (Dixon, J., Bell, K., 2020) to increase the capacity of batteries and shift the time of charge of electric vehicles to avoid peak loads on the net. It was proposed in (Fathabadi, H.,

2020) to ensure the stability of the power net with the use of additional accumulation and wind energy. The search for maximum power with a variable step which is applied to both the photoelectric and wind part of the station was used. Moreover, it was proposed to use an auxiliary power source with control of redistribution of produced and consumed energy.

The system produces additional electricity when the output of photoelectric and wind energy is less than that required for charging. The system electrolyzer produces hydrogen by absorbing additional electrical power available in the system when the production of photoelectric and wind energy exceeds the charging requirement. Thus, an additional energy system acts as a storage tank adjusting the charge power according to energy consumption. The use of an additional power source and an additional storage capacity relates to the lack of assessment of change in the battery capacity as an integral part of the charging station. In conditions of distributed generation of electric energy, the operation of the battery capacity as an integral part of the circuit design of charging stations becomes fundamental in terms of voltage regulation in the distribution system. The neural model of predicting changes in parameters of the electrical system based on distributed parameters (Jia, Y., Liu, X. J., 2014) does not estimate the change in battery capacity in terms of matching power production and consumption.

The presented analysis of literature allows us to assess optimization of the charging stations based on economic and environmental principles of connection to renewable energy sources. Control of electricity consumption and production is carried out with the use of additional devices for voltage regulation in the distribution system, increasing the capacity of batteries, or inclusion of additional storage devices which requires additional costs. The rechargeable battery as a mandatory element of the technological scheme of the photoelectric charging station can become the main element of voltage regulation in the distribution system. This is possible if changes in its capacity are predicted. In this case, the battery becomes the basis for redistribution of electric energy between the network and the photoelectric module, i.e., it becomes a voltage regulator in the distribution system. Moreover, the assessment of change in the battery capacity makes it possible to maintain the power factor of the photoelectric charging station.

Therefore, it was proposed to measure the voltage at the input to the hybrid inverter and in the distribution system to assess their ratio. Making advanced decisions to change the level of power transmission to the network will enable the regulation of voltage in the distribution system to match the production and consumption of energy. The above substantiates the need for further studies in this area.

Methodological and Mathematical Substantiation

Distributed generation of electric energy using renewable sources requires connection to Smart Grid technologies for integration into the electric network. For example, author's work (Chaikovskaya, E., 2017) based on predicting voltage changes in a storage battery is devoted to connection to Smart Grid technologies. A technology of maintaining change in the capacity of a storage battery during the measurement of electrolyte temperature in a set of accumulators was presented. The use of an integrated system of estimating a change in voltage based on matching electrochemical and diffusion processes of charge and discharge enables making advanced decisions on boosting to prevent impermissible overcharge and discharge.

The author's work (Chaikovskaya, E., 2019) tackles forecasting change in the battery capacity for connection to Smart Grid technologies. It presents an integrated system of maintaining the operation of a wind-solar electrical system. Making advanced decisions on changing the power of a thermoelectric accumulator is based on establishing a ratio between the voltage measured at the input to a hybrid charge controller and at an inverter output when measuring frequency. Up to 30% reduction of thermoelectric accumulator charge time was provided based on a change of the number of turns of a circulating pump motor to change flow rate and temperature of heated water.

There are devices for charging electric vehicles which differ from each other by the type of current used and charging time. For example, Mode 3 of a charging station operation using alternating current makes it possible to charge electric cars of medium power in 4 hours using a 10-kW charger. Fast charge in Mode 4 using direct current restores the capacity of electric car batteries to 80% in half an hour. A serious complication occurring in the use of AC charging stations consists of a risk of peak loading of mains. In conditions of the growing number of electric vehicles and irregularity of charge, there is a need to build charging stations using renewable energy sources.

Maintaining of power factor of the photoelectric charging station as regards redistribution of produced and consumed electricity is an urgent problem of further development of Smart Grid technologies. To this end, it is necessary to predict changes in battery capacity and power factor of the photoelectric charging station when measuring the input voltage of a hybrid inverter and voltage in the distribution system to assess their ratio. Advanced decisions concerning change in the level of power transmission to the electric network make it possible to adjust the voltage in the distribution system to maintain the balance of active and reactive power without the use of additional devices. The power factor of photoelectric charging stations is maintained by coordination of production and consumption of electric energy. This enables the prevention of peak loads in the electric network in conditions of satisfaction of growing consumer demands.

One of the main properties of energy systems is the mandatory exchange of substance, energy and information with the environment.

Thus, the photoelectric charging station – open integrated dynamic system, the operation of which requires to predict changes in battery capacity and power factor of the photoelectric charging station when measuring the input voltage of a hybrid inverter and voltage in the distribution system to assess their ratio. The dynamic characteristics of the photoelectric charging station with sufficient accuracy for practice can be described by a finite set of parameters for changes in time, and the spatial coordinate that coincides with the direction of flow of the medium, such as changes battery capacity and power factor of the photoelectric charging station.

Therefore, the dynamic description of the photoelectric charging station most fully and multifacetedly characterizes its operation. Thus, it is possible to determine that the real photoelectric charging station is a dynamic system, the mathematical model of which reflects the properties of the transformation of influences, i.e., its dynamic properties. Due to the fact that the photoelectric charging station reflects the dynamic peculiarities due to the nature of reactions to influences, the supporting of the functioning photoelectric charging station should be part of such a technological system, which is based on a dynamic system. Based on the methodological, mathematical, logical substantiation of the technological systems (Chapter 1) the architecture, mathematical substantiation of the architecture (1), mathematical substantiation of maintenance of the operation (2) of the photoelectric charging station Smart Grid are proposed (Figure 6.1).

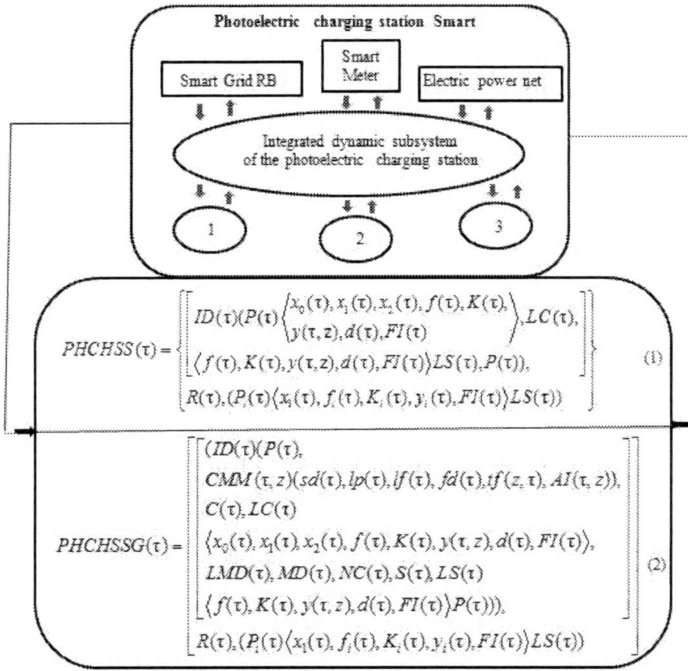

Figure 6.1. Photoelectric charging station Smart Grid: the architecture: RB – rechargeable battery; Smart Meter is a two-way counter of changes in the level of power transmission to the network; 1 – the charging unit; 2 – the discharging unit; 3 – the unit of assessing the functional efficiency. Mathematical substantiation of the architecture (1). Mathematical substantiation of maintenance of the operation (2).

A photoelectric charging station is a dynamic system, the operation of which is the reproduction of a change in external, internal influences and initial conditions, for example, changes in solar radiation, power consumption for charging electric vehicles, a voltage in the distribution system, etc. Therefore, when designing a photoelectric charging station, an integrated dynamic subsystem is laid down in its base (Figure 6.1). That is why, when designing. That is why, when designing a photoelectric charging station, underlying which is an integrated dynamic subsystem (Figure 6.1). The integrated dynamic subsystem includes the following components: mains, photoelectric solar panels, a hybrid inverter, rechargeable batteries, a two-way Smart Meter counter and a charger. When representing the system design as the organization of a complex system, it was expanded by building up the dynamic subsystem blocks that forecast the process components around its base. Other components of the technological system include the units of charge and discharge and functional efficiency estimation in a coordinated interaction with the dynamic subsystem (Figure 6.1).

The mathematical substantiation of the architecture of the photoelectric charging station Smart (1), (Figure 6.1), based on the methodology of the mathematical description of dynamics of power systems, the method of the graph of cause-effect relations (Chapter 1) is proposed.

Where *PHCHSS* (τ) – photoelectric charging station Smart; τ – time, s; *ID*(τ) – integrated dynamic subsystem (mains, a photoelectric module, a hybrid inverter, a rechargeable battery, a two-way Smart Meter counter and a charger); *P*(τ) – properties of the components of the photoelectric charging station; *x*(τ) – impacts (changes in solar radiation, power consumption for charging electric vehicles, voltage in the distribution system, etc.; *f*(τ) – parameters that are measured: (voltage at the input to the hybrid inverter, voltage in the distribution system); *K*(τ) – coefficients of mathematical description of dynamics of change in the capacity of the storage battery, power factor of the photoelectric charging station; *y*(τ, *z*) – predicted output parameters (battery capacity, power factor of the photoelectric charging station) *z* – coordinate of the length of the battery plates, m; *d*(τ) – dynamic parameters (battery capacity, power factor of the photoelectric charging station); *FI*(τ) – functional resulting information on decision making; *LC*(τ) – logical relations regarding the control of the photoelectric charging station workability; *LS*(τ) – logical relations regarding the identification of the state of the photoelectric charging station; *R*(τ) – logical relations in *PHCHSS* (τ) to confirm the correctness of decisions made from the units of the photoelectric charging station. Indices: *i* – the number of elements of the photoelectric charging station; 0, 1, 2 – initial stationary mode, external, internal nature of impacts.

The mathematical substantiation of maintenance of the operation of the photoelectric charging station Smart Grid (2), (Figure 6.1), based on the methodology of the mathematical description of dynamics of power systems, the method of the graph of cause-effect relations [Chapter 1] is proposed. The basis of the proposed rationale is the mathematical description of the architecture of the photoelectric charging station Smart (1), (Figure 6.1). Prediction of changes in the battery capacity and the power factor of the photoelectric charging station enables making advanced decisions to change the level of power transmission to the network in order to maintain voltage in the distribution system. The change in the ratio of voltage at the input to the hybrid inverter and voltage in the distribution system is assessed.

The mathematical substantiation of Smart Grid maintenance of the operation of the photoelectric charging station (2) is proposed (Figure 6. 1).

Where *PHCHSSG* (τ) – Smart Grid maintenance of the operation of the photoelectric charging station; τ – time, s; *ID*(τ) – integrated dynamic subsystem (mains, a photoelectric module, a hybrid inverter, a rechargeable battery, a two-way Smart Meter counter and a charger). *P*(τ) – the properties of the elements of the integrated dynamic subsystem, units of the photoelectric charging station; *CMM*(τ, *z*) – complex mathematical modeling of the dynamics of changes in the battery capacity, power factor of the photoelectric charging station; *sd*(τ) – the input data **(**the power of photoelectric module and charge of electric cars, the rechargeable battery and its type and capacity, the two-way Smart Meter counter and its type and the charger and its type; *lp*(τ) – the boundary change in parameters (the voltage at the input to the hybrid inverter and voltage in the distribution system**;** *lf*(τ) – the levels of operation of the photoelectric charging station**;** *fd* (τ) – the obtained parameters **(**mode parameters of the photoelectric charging station)**;** *tf*(τ,*z*) – the transfer function of predicted parameters – in the battery capacity, power factor of the photoelectric charging station**;** *AI*(τ,*z*) – the standard information regarding the evaluation of the maximum admissible change in the battery capacity, power factor of the photoelectric charging station; *C*(τ) – the control of workability of the photoelectric charging station; *LC*(τ) – the logical relations of the control of the photoelectric charging station workability; *x*(τ) – impacts (changes in solar radiation, power consumption for charging electric vehicles, voltage in the distribution system, etc; *f*(τ) – the measured parameters: (the voltage at the input to the hybrid inverter and voltage in the distribution system are measured to assess their ratio); *K*(τ) – the coefficients of the mathematical description of the dynamics of a change in the coordinate of the length of the battery plates; *y*(τ, *z*) – the output parameters (the battery capacity, power factor of the photoelectric charging station); *z* – the coordinate of the length of the battery plates, m; *d*(τ) – the dynamic parameters of estimation of a change in the battery capacity, power factor of the photoelectric charging station; *FI*(τ) – functional resulting information on decision making; *LMD*(τ) – the logical relations of decision making; *MD*(τ) – decision making; *NC*(τ) – the new conditions of the photoelectric charging station operation; *S*(τ) – the identification of the state of the photoelectric charging station; *LS*(τ) – the logical relations of identification of the state of the cogeneration system; *R*(τ) – the logical relations between the dynamic subsystem and units of charge, discharge, functional estimation of efficiency that belong to the cogeneration system. Indices: i – the number of elements of *BDCSSG* (τ); 0, 1, 2 – the initial, external, and internal character of influences.

Mathematical substantiation of the architecture of the photoelectric charging station Smart (1) and mathematical substantiation of maintenance of the operation of the photoelectric charging station Smart Grid (2) (Figure 6.1) make it possible to maintain the operation of the photoelectric charging station using the following actions:

- Workability control ($C(\tau)$) of the dynamic subsystem based on complex mathematical ($CMM(\tau, z)$) and logical ($LC(\tau)$) modeling regarding obtaining standard ($AI(\tau,z)$) estimate of a change in the battery capacity, power factor of the photoelectric charging station;
- Workability control ($C(\tau)$) of the dynamic system based on complex mathematical ($CMM(\tau, z)$) and logical ($LC(\tau)$) modeling regarding the obtaining functional (FI (τ)) estimate of a change in the battery capacity, power factor of the photoelectric charging station;
- Decision making ($MD(\tau)$) with the use of the functional resulting information (FI (τ)), obtained based on logical modeling ($LMD(\tau)$); decision making to change the level of power transmission to the network to maintain the power factor of the photoelectric charging station;
- Identification ($S(\tau)$) of the new conditions of functioning of the photoelectric charging station ($NC(\tau)$) based on logical modeling ($LS(\tau)$) as a part of the dynamic subsystem and confirmation of new operating conditions based on logical modeling ($R(\tau)$) from the units of the photoelectric charging station.

Maintaining of Voltage in the Distribution System Based on a Prediction of Changes in the Battery Capacity

According to Formulas (1), (2), the prediction of a change in the storage battery capacity and power factor of the photoelectric charging station are proposed. The voltage at the input to the hybrid inverter and voltage in the distribution system are measured to assess their ratio. Transfer functions by channels: "battery capacity – voltage at the input to the hybrid inverter," "power factor of the photoelectric charging station – voltage in the distribution system," which was obtained as a result of solving a system of nonlinear differential equations, are presented as follows:

$$W_{CE-U_1} = \frac{K_{ce}K}{(T_e S+1)\beta - 1}\left(1-e^{-\gamma\xi}\right), \tag{3}$$

$$W_{pf-U_2} = \frac{K_{pf}K}{(T_e S+1)\beta - 1}\left(1-e^{-\gamma\xi}\right), \tag{4}$$

where

$$K_{ce} = \frac{I_1 U_1}{(U_1 - U_2)};\; K_{pf} = \frac{I_2(U_1 - U_2)}{N};\; K = \frac{m(\theta_0 - \sigma_0)}{G_0};$$

$$L = \frac{G_e C_e}{\alpha_0 h_0};\; T_e = \frac{g_e C_e}{\alpha_0 h_0};\; \beta = T_m S + \varepsilon + 1;\; T_m = \frac{g_m C_m}{\alpha_0 h_0};$$

$$\varepsilon = (1-L);\; \gamma = \frac{(T_e S+1)\beta - 1}{\beta};\; \xi = \frac{z}{L},$$

where CE is the battery capacity, Ah; PF – power factor of the photoelectric charging station; I_1, I_2 – currents at the input of the hybrid inverter, in the distribution network, respectively, A; U_1, U_2 – voltages at the input to the hybrid inverter and in the distribution system, respectively, V; N – the power of the photoelectric charging station, kW; C is the specific thermal capacity, kJ/(kg·K); α is the heat transfer factor, kW/(m^2·K); G – consumption of substance, kg/s;g – specific weight of a substance, kg/m^3; h is the specific surface, m^2/m; σ, θ – the temperature of electrolyte at the battery output and at the distribution wall, respectively, K; z – the coordinate of the length of the battery plates, m; $T_{e,}$, T_m – time constants that characterize thermal storage capacity of electrolyte and metal, s; m is the indicator of the dependence of heat transfer factor on consumption; ι – time, s; S – the Laplace transform parameter; $S=\omega j$; ω: frequency, 1/s. Indices: 0 – initial stationary mode; e – electrolyte; m: metal wall.

Transfer functions by channels: "battery capacity – voltage at the input of the hybrid inverter," "power factor of the photoelectric charging station – voltage in the distribution system" were obtained by solving a system of nonlinear differential equations using the Laplace transform. The systems of differential equations include an equation of state as an estimate of the physical model of the photoelectric charging station, an equation of energy of the battery charge and discharge, an equation of heat balance for the wall of the battery plates.

The equation of charge and discharge energies was composed with the representation of a change in electrolyte temperature in pores of the plates and above the plates both in time and along the spatial coordinate of the battery plates. The transfer functions include coefficients K_{ce}, K_{pf} which estimate change in the battery capacity and the power factor of the photoelectric charging station. When analyzing the obtained mathematical model, internal parameters to be diagnosed as being a part of coefficients of the equations of dynamics K_{ce}, K_{pf} were established. In real conditions of operation of the power system at a transition from stationary states and under external and internal effects, reorganization of the coefficients of equations of dynamics in time occurs because of a change in the diagnosed internal parameters. A real part of the transfer functions are separated:

$$O(\omega)=\frac{(L_1A_1)+(M_1B_1)(1-L)}{\left(A_1^2+B_1^2\right)}. \tag{5}$$

The K factor includes the temperature of the separating wall θ:

$$\theta=\left(\alpha_{\text{edch}}\left(\sigma_1+\sigma_2\right)/2\right)+\left(A\left(t_1+t_2\right)/2\right)/\left(\alpha_{\text{edch}}+A\right), \tag{6}$$

where σ_1, σ_2 are the temperature of electrolyte at the inlet and outlet of the battery, K, respectively; t_1, t_2 are the temperature of electrolyte in pores of the plates and above the plates at the battery inlet and outlet, respectively, K; α is the heat transfer factor, kW/(m^2·K). Indices – e – electrolyte; dch–discharge.

$$A=1/\left(\delta_{\text{m}}/\lambda_{\text{m}}+1/\alpha_{ech}\right), \tag{7}$$

where δ is the battery plate wall thickness, m; λ is the thermal conductivity of metal of the battery plate, kW/(m·K). Indices: m is the metal wall of the battery plate; ch – charge.

.

To use the real part $O(\omega)$, the following factors were obtained:

$$A_1 = \varepsilon - T_e T_m \omega^2; \quad (8)$$

$$A_2 = \varepsilon + 1; \quad (9)$$

$$B_1 = T_e \varepsilon \omega + T_e \omega + T_m \omega; \quad (10)$$

$$B_2 = T_m \omega; \quad (11)$$

$$C_1 = \frac{A_1 A_2 + B_1 B_2}{A_2^{\;2} + B_2^{\;2}}; \quad (12)$$

$$D_1 = \frac{A_2 B_1 - A_1 B_2}{A_2^{\;2} + B_2^{\;2}}; \quad (13)$$

$$L_1 = 1 - e^{-\xi C_1} \cos\left(-\xi D_1\right); \quad (14)$$

$$M_1 = -e^{-\xi C_1} \sin\left(-\xi D_1\right). \quad (15)$$

The transfer functions (3), (4) which was obtained based on the use of the operator method of solving the system of nonlinear differential equations, includes the Laplace transform parameter – S (S = ωj), where ω is the frequency, 1/s. To switch from the frequency area to the time area, a real part (5), obtained as a result of the mathematical treatment of transfer functions, was separated.

It is this part that is included in the integrals (16), (17) which makes it possible to obtain dynamic characteristics of a change the battery capacity, power factor of the photoelectric charging station using the inverse Fourier transform:

$$CE(\tau) = \frac{1}{2\pi}\int_0^{\infty} K_{ce} KO(\omega)\sin(\tau\omega/\omega)d\omega, \tag{16}$$

$$PF(\tau) = \frac{1}{2\pi}\int_0^{\infty} K_{pf} KO(\omega)\sin(\tau\omega/\omega)d\omega, \tag{17}$$

where CE is the battery capacity, Ah; PF is the power factor of the photoelectric charging station.

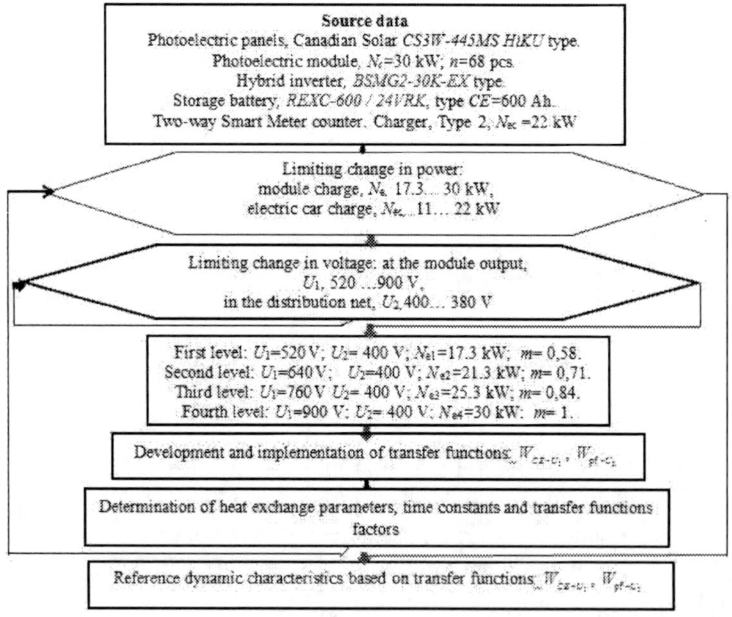

Figure 6.2. Block diagram of comprehensive mathematical modeling of the photoelectric charging station: N_e, N_{ec} – the power of photoelectric module and charge of electric cars, respectively, kW; CE– battery capacity, Ah; U_1, U_2 – voltage at the input to the hybrid inverter and in the distribution system, respectively, V;

n – the number of photoelectric panels; m – the power level of the photoelectric charging station.

So, for obtaining of reference estimation of change in the battery capacity and power factor of the photoelectric charging station, a block diagram is proposed (Figure 6.2) with, for example, the initial data of a photoelectric charging station with a power of 30 kW, for charging electric vehicles with a power of 22 kW.

The following levels of operation of the photoelectric charging station have been established for the change in the voltage at the input to the hybrid inverter and in the distribution system: – first level: 520–400V; second level: 640–400V; third level; fifth level: 900–400V. They correspond to changes in the power of the photoelectric module: 17.3, 21.3, 25.3, 30, respectively, kW and the power level: 0.58, 0.71, 0.84, 1.

Table 6.1. Mode parameters of the photoelectric charging station

Levels of operation	N_e, kW	U_1, V	U_2, V	m
First level	17.3	520	400	0.58
Second level	21.3	640	400	0.71
Third level	25.3	760	400	0.84
Fourth	30	900	400	1

Note: N_e – the power of photoelectric charging station, kW; U_1, U_2 – voltages at the input to the hybrid inverter and in the distribution system, respectively, V; m – power level of the photoelectric charging station.

Table 6.2. Heat exchange parameters of the battery

Operation levels	Parameter		
	α_{ch}, W/(m²·K)	α_{dch}, W/(m²·K)	k, W/(m²·K)
Charge, discharge	15.298	15.232	3.359

Note: α_{ch} – coefficient of heat transfer from the electrolyte to a wall of the battery plate when charged, W/(m²·K); α_{dch} – coefficient of heat transfer from the wall of the battery plate to the electrolyte when discharged, W/(m²·K); k – coefficient of heat exchange, W/(m²·K).

Table 6.3. Time constants and coefficients of mathematical models of dynamics of the photoelectric charging station

Operation levels	T_e, s	T_m, s	ε	ζ	L, m
Charge	1467.56	13352.5	0.973	0.647	35.51
Discharge	1466.94	13346.86	0.973	0.647	35.51

According to Formulas (1) - (4) and the proposed block diagram (Figure 6.2), the results of reference information obtained on the basis of complex mathematical modeling of the photoelectric charging station Smart Grid is presented (Tables 6.1 - 6.4).

Time constants and the coefficients that are components of mathematical models of dynamics (3), (4) presented in Table 6.3 were obtained based on the parameters of heat exchange for charge and discharge of the battery presented in Tables 6.1, 6.2.

Based on the proposed mathematical substantiation Smart Grid maintenance of functioning of the photoelectric charging station (1) to (4) the block diagram for the control of serviceability of the photoelectric charging station (Figure 6.3) is developed.

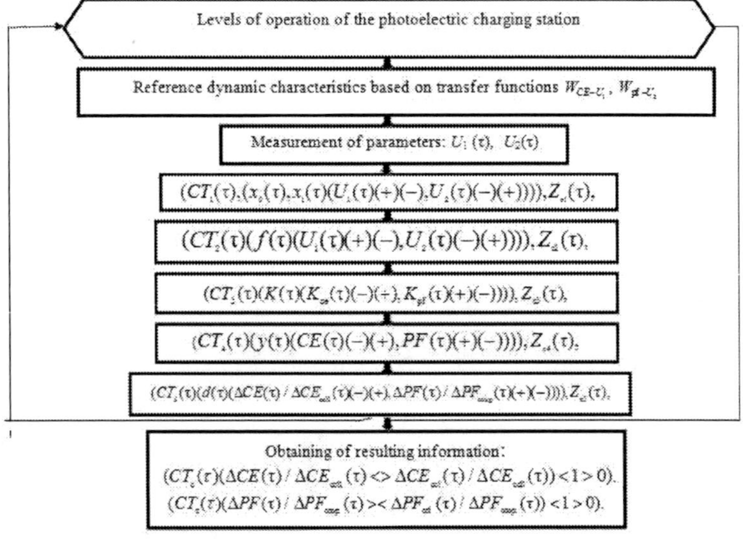

Figure 6.3. Block diagram of the photoelectric charging station functioning control: U_1, U_2 – voltages at the input to the hybrid inverter and in the distribution system, respectively, V; *CE* – battery capacity, Ah; *KF* – power factor of the photoelectric charging station; *CT* – event control; *Z* – logical relations; *d* – dynamic parameters; *x* – effects; *f* – parameters measured; *y* – parameters predicted; *K* – coefficients of mathematical description; ι – time, s. Indices: c – control of operability; ccup, ccll – constant calculated value of the parameter of upper, lower levels of operation, respectively; ccl – constant calculated value of the level of operation parameter; 0, 1, 2 – initial stationary mode, external, internal influences; 3 – coefficients of dynamics equations; 4 – significant predicted parameters; 5 – dynamic parameters.

Control of workability of the photoelectric charging station (Figure 6.3) enables obtaining the resulting information on decision-making about the maintenance of the voltage in the distribution system.

Based on the proposed mathematical substantiation Smart Grid (1) to (4), (Figure 6.3) the block diagram of maintenance of operation of the photoelectric charging station has been developed based on maintaining the distribution system voltage (Figure 6.4) is developed.

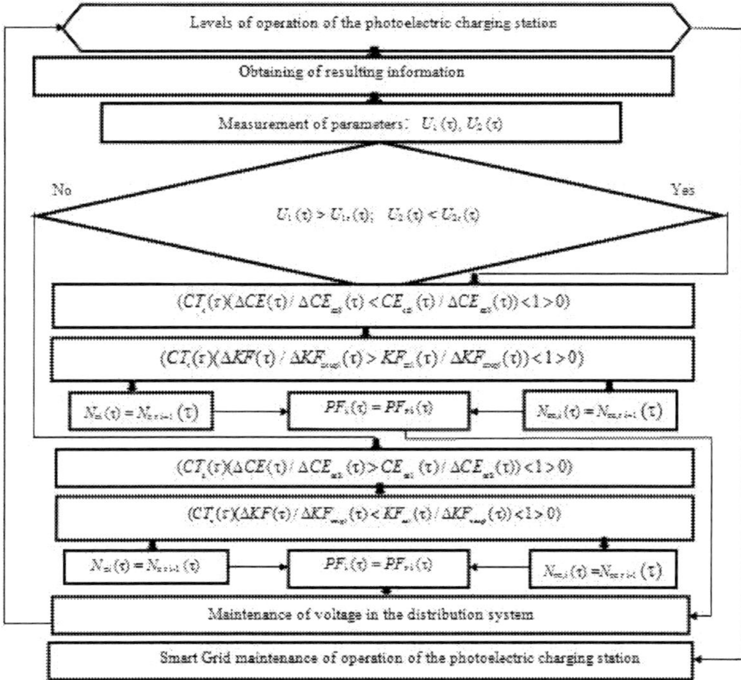

Figure 6.4. Block diagram of maintenance of operation of the photoelectric charging station: U_1, U_2 – voltage at the input to the hybrid inverter and in the distribution system, respectively, V; *CE* – battery capacity, Ah; *KF* – power factor of the photoelectric charging station; N_e, N_{ec} – power of the photoelectric module and charge of electric cars, respectively, kW; ι – time, s; Indices: *i* – number of operation levels; r – reference value of the parameter; ccupl, ccll – constant calculated value of the parameter of the upper and lower levels of operation, respectively; ccl – constant calculated value of the parameter of the operation level.

Voltage maintenance in the distribution system (Figure 6.4) makes it possible to ensure the operation of the photoelectric charging station.

Table 6.4. Integrated system of the charging station operation maintenance

Time, τ, 10^3 s	Change in parameters	$\Delta CE(\tau)/\Delta CE_1(\tau)$	$CE(\tau)$, Ah	$\Delta PF(\tau)/\Delta PF(\tau)_2$	$PF(\tau)$
0	Charge – discharge U_1=520 V; U_2=400 V; N_e=17.3 kW; m=0.58	1	600	0.2400	0.6760
3	Charge – discharge U_1=544 V; U_2=395 V; N_e=18.1 kW; m=0.6	0.8425	576.38	0.30	0.7000
6	Charge – discharge U_1=568 V; U_2=390 V; N_e=18.9 kW; m=0.63	0.7364	560.47	0.3560	0.7224
9	Charge – discharge U_1=592 V; U_2=385 V; N_e=19.7 kW; m=0.66	0.66	549.01	0.4140	0.7456
12	Charge – discharge U_1=616 V; U_2=380 V; N_e=20.5 kW; m=0.68	0.6023	540.35	0.4720	0.7688
15	Decision making m=0.71; U_1=640 V; U_2=400 V; N_e=21.3 kW	0.6154	542.31	0.48	0.7720
18	Charge – discharge U_1=664 V; U_2=395 V; N_e=22.1 kW; m=0.74	0.5696	535.45	0.5380	0.7952
21	Charge – discharge U_1=688 V;U_2=390 V; N_e=22.9 kW; m=0.76	0.5328	529.93	0.5960	0.8184
24	Charge – discharge U_1=712 V; U_2=385 V; N_e=23.7 kW; m=0.79	0.5025	525.39	0.6540	0.8416
27	Charge – discharge U_1=736 V; U_2=380 **V**; N_e=24.5 kW; m=0.82	0.4771	518.58	0.7120	0.8648
30	Decision making m=0.84; U_1=760 V; U_2=400 V; N_e=25.3 kW	0.4872	520.085	0.72	0.8680

Time, τ, 10^3 s	Change in parameters	$\Delta CE(\tau)/\Delta CE_1(\tau)$	$CE(\tau)$, Ah	$\Delta PF(\tau)/\Delta PF(\tau)_2$	$PF(\tau)$
33	Charge – discharge U_1=784 V;U_2=395 V; N_e=26.1 kW; m=0.87	0.4450	518.25	0.7779	0.8912
36	Charge – discharge U_1=810 V; U_2=390 V; N_e=27 kW; m=0.9	0.4286	516	0.84	0.9160
39	Charge – discharge U_1=834 V; U_2=385 V; N_e=27.8 kW; m=0.93	0.4142	513.84	0.8980	0.9392
42	Charge – discharge U_1=858 V; U_2=380 V; N_e=28.6 kW; m=0.95	0.4154	514.02	0.9559	0.9624
45	Decision making m=1; U_1=900 V; U_2=400 V; N_e=30 kW	0.4341	511.22	1	0.98

Note: U_1, U_2 – voltage at the input to the hybrid inverter and in the distribution system, respectively, V; CE – battery capacity, A·h; KF – power factor of the photoelectric charging station; N_e – power of the photoelectric module, kW; m – level of power of the photoelectric charging station; ι – time, Indices– i – number of operation levels; 1, 2 – constant calculated value of the parameter of lower and upper levels of operation, respectively.

Results and Discussion

The Smart Grid System of Maintaining the Operation of the Photoelectric Charging Station at the Decision-Making Level

A comprehensive integrated system has been developed (Table 6.4) for maintaining the operation of the photoelectric charging station based on a prediction of changes in the battery capacity and power factor of the photoelectric charging station.

Advanced decisions on change in the level of transmission of electrical energy to the network make it possible to maintain voltage in the distribution system through maintaining the power factor of the photoelectric charging station. Continuous measurement of the voltage at the input to the hybrid inverter and in the distribution system takes place to assess their ratio.

The integrated Smart Grid system of maintenance of operation of the photoelectric charging station (Table 6.4) provides an opportunity to coordinate electric power production and consumption.

Coordination of Electric Power Production and Consumption Based on Voltage Maintenance in the Distribution System

The battery capacity at a specified time point was determined as follows:

$$CE_{i+1}(\tau) = CE_i - \\ + \begin{pmatrix} \Delta CF_i(\tau) / \Delta CE_{\text{ccll}}(\tau) - \\ -\Delta CE_{i+1}(\tau) / \Delta CE_{\text{ccll}}(\tau) \end{pmatrix} (CE_2 - CE_1), \tag{18}$$

where CE – the battery capacity, Ah; CE_1, CE_2 – initial and final values of the battery capacity, Ah; τ – time, s. Indices– ccll – constant calculated value of the parameter of lower level of operation; i – the number of levels of operation of the photoelectric charging station.

The power factor of the photoelectric charging station at the set time is determined as follows:

$$PF_{i+1}(\tau) = PF_i + \\ + \begin{pmatrix} \Delta PF_{i+1}(\tau) / \Delta PF_{ccupl}(\tau) - \\ -\Delta PF_i(\tau) / \Delta PF_{ccupl}(\tau) \end{pmatrix} (PF_2 - PF_1), \tag{19}$$

where PF – power factor of the photoelectric charging station; PF_1, PF_2 – initial and final values of the power factor; τ – time, s. Indices – ccupl – constant calculated value of the parameter of the upper level of operation; i – the number of levels of operation of the photoelectric charging station.

For example, in a period of $15 \cdot 10^3$ s (4.17 hrs), the battery capacity was predicted to increase to the level of 542.31 Ah with voltage growth at the input to the hybrid inverter at the level of 540 V. Value of the battery capacity was determined using the formula (18) as follows (Table 6.4, Figure 6.5):

542.31 Ah=540.35+(0.6154–0.6023) (600–450).

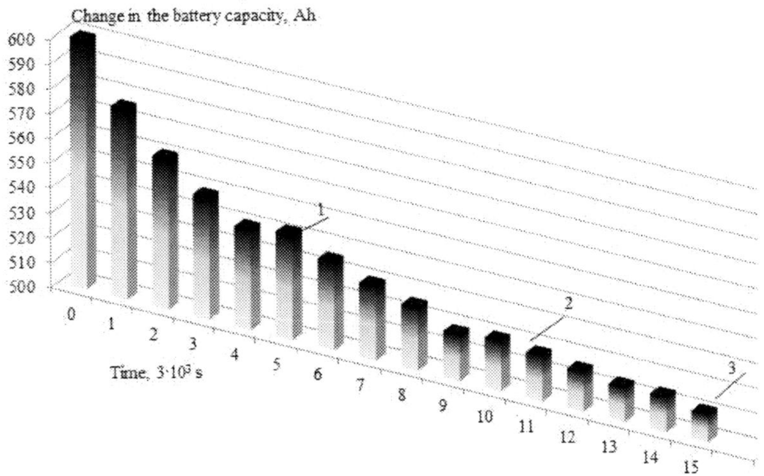

Figure 6.5. Maintenance of change in the battery capacity. 1, 2, 3 are the points where decisions were made to change the level of power transmission to the network.

During this period, it was necessary to make an advanced decision to raise the level of power transmission to the network from 0.68 to 0.71. The voltage level in the distribution system was set at 400 V and the power factor of the photoelectric charging station was at 0.7720.

The value of the power factor in this period was determined as follows using formula (19) (Table 6.4, Figure 6.6):

$$0.7720 = 0.7688 + (0.48 - 0.4720)\ (0.98 - 0.58).$$

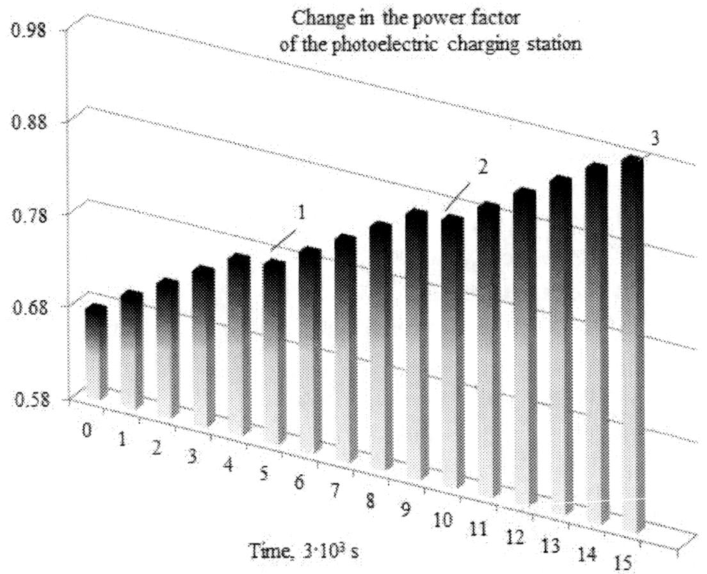

Figure 6.6. Maintenance of operation of the photoelectric charging station. 1, 2, 3 are the points where decisions were made on the change of level of power transmitted to the network.

Performing such actions will enable maintenance of voltage in the distribution system to coordinate the production and consumption of electric power.

Conclusion

1. The mathematical substantiation of maintenance of the operation of the technological systems Smart Grid based on the architecture and mathematical description of the architecture of the technological systems, methodology of the mathematical description of dynamics of power systems, the method of the graph of cause-effect relations are proposed. The mathematical substantiation is based on a systematic approach to complex mathematical and logical modeling in the technological system as a dynamic system. Decision-making to support the functioning of energy systems is based on a mathematical description of the architecture of technological systems, the methodology of mathematical description of the dynamics of energy systems, the method of the graph of causation. Coordination of energy production and consumption is based on forecasting changes in the parameters of technological processes.
2. The integrated Smart Grid system of harmonization of production and consumption of electric power and heat in the biogas cogeneration system with the use of heat-pumping power supply of the biogas plant, which uses fermented wort as a low-potential source of power, is proposed. A change in the power factor of the cogeneration system, the temperature of local water is predicted by measuring the voltage at the inlet to the inverter, at the outlet from the inverter and voltage frequency. In the engine cooling circuit, the temperature of cooling water at the inlet to the heat exchanger, at the outlet from the heat exchanger, and the return water temperature are measured. It was proposed to estimate a change in the ratio of voltage at the inlet to the inverter and at the outlet from the inverter. Making forestalling decisions to change the power of the heat pump and the number of plates in the heat exchanger of the engine cooling circuit makes it possible to maintain the voltage at the entrance to the inverter and the temperature of the heated local water. Functional estimation of a change in power factor of the cogeneration system in the range of 85–95%, temperature of local water in the range of 30–55°C at the compensation of reactive power of up to 40% was obtained.

3. The integrated Smart Grid System of harmonization of production and consumption of electric power and heat in the biodiesel cogeneration system is proposed. A change in the ratio of production and consumption of electric power and heat, change in the temperature of oil and change in the temperature of local water of the engine cooling circuit is predicted by measuring temperature of biodiesel at the outlet from the heat exchanger, the temperature of cooling water at the outlet of the heat exchanger the engine cooling circuit and the temperature of return water Making forestalling decisions to support the supply of fresh oil for heating, to support the supply of heated oil for etherification, changing number of plates of the biodiesel heat exchanger and the changing number of plates of the heat exchanger of the engine cooling circuit makes it possible to maintain the ratio of production and consumption of electric power and heat and the temperature of the heated oil and local water. Functional estimation of a change in the ratio of production and consumption of electric power and heat of the cogeneration system in the range of 0.6606 – 0.6611, temperature of oil in the range of 20–45°C, temperature of local water in the range of 30–55°C was obtained. Determining final functional information provides an opportunity to make forestalling decisions to changing the ratio of production and consumption of electric power and heat to maintain the functioning of the cogeneration system.
4. The integrated Smart Grid System of harmonization of production and consumption of electric power and heat in the pellet cogeneration system is proposed. A change in the ratio of production and consumption of electric power and heat, change in the moisture content of the air in the drying chamber and change in the temperature of local water of the second circuit of the cogeneration system is predicted by measuring temperature of air at the outlet from the heat exchanger, the temperature of the of flue gases at the inlet of the heat exchanger and the return water temperature. Making forestalling decisions to support the supply of raw timber for drying, dried timber for production of the pellet fuel, changing number of the number of revolutions of the electric motor of the air fan and the changing number of plates of the heat exchanger of the second circuit makes it possible to maintain the ratio of production and consumption of electric power and heat and the moisture content of the air and local water. Functional estimation of a change in the ratio of production

and consumption of electric power and heat of the cogeneration system in the range of 0.6294 – 0.6305, moisture content of the air in the range of 12–40%, temperature of local water in the range of 90–79.83°C was obtained. Determining final functional information provides an opportunity to make forestalling decisions to changing the ratio of production and consumption of electric power and heat to maintain the functioning of the cogeneration system.

5. Integrated Smart Grid Systems of harmonization of production and consumption of electric power of the heat pump power supply and hot water power supply in the wind-solar electric system with hybrid solar collectors are developed are developed. Integrated systems based on predicting changes in the power factor, temperature of local water when measuring voltage from hybrid solar collectors at the input to the grid inverter, voltage at the output of the frequency converter and voltage frequency. The adoption of advanced decisions to maintain the temperature of local water by changing the power of the electric motor of the heat pump compressors and electric motor of the circulation pump based on establishing the ratio of the voltage at the input to the grid inverter and the voltage at the output of the frequency converter are measured. The power factor of the wind-solar electric system is maintained. Functional estimation of a change in power factor of the wind-solar electric system in the range of 58–98%, temperature of local water in the range of 30–55°C for hot water power supply and in the range of 35–55°C for heat pump power supply. Determining final functional information provides an opportunity to make forestalling decisions on a change in the of the electric motor of the heat pump compressors and electric motor of the circulation pump to prevent the peak load of the power system under voltage regulation conditions when connecting the heat pump power supply and hot water supply.
6. Integrated Smart Grid System of harmonization of production and consumption of electric power based on a prediction of changes in the battery capacity in photoelectric charging station is developed. Advanced decisions on the change in power transmission capacity have made it possible to regulate voltage in the distribution system by maintaining the power factor of the photoelectric charging station. Voltages at the input to the hybrid inverter and in the dis-tribution system were measured to assess their ratio. A functional estimate of changes in the battery capacity and power factor of the photoelectric

charging station was obtained. Maintenance of voltage in the distribution system was realized based on resulting operation data to estimate a change in the battery capacity. Advanced decision-making has made it possible to raise the power factor of the photoelectric charging station up to 40% due to matching the electric power production and consumption. Maintenance of operation of the photoelectric charging station using the developed Smart Grid technology has enabled prevention of peak loading of the power system due to a 20% reduction of power consumption from the network.

References

Chaikovska, Ye. Ye. (2004). Dopusk yak struktura. [*Admission as a structure*]. *Bulletin of the Lvov Polytechnic National University. Thermal energy. Environmental engineering. Automation.* Lvov. 506. 285-290.

Chaikovska, Ye. Ye. (2016). Informatsiyni tekhnolohiyi pidtrymky funktsionuvannia enerhetychnykh system na rivni pryiniattia rishen. [*Information technologies supporting the functioning of energy systems at the level of initiation of decisions*]. *Computer science Culture. Equipment: coll. theses add. IV Ukrainian-Nim. Conf. Odessa*, 32–33.

Chaikovska, Ye. Ye. (2017). Intelectualni systemi pidtrymky functcionuvania energetichnih system na rivni priniattia richen [*Intelligent systems supporting the functioning of energy systems at the level of priniattia are decided*]. *Enerhetyca. economica, tehnologii, ekologia,* 3, 114–118. https://doi.org/10.20535/1813-5420.3.2017.117377.

Chaikovskaya, E. (2014). Maintaining the relation between production and consumption of electricity and heat at decision-making level. *Eastern-European Journal of Enterprise Technologies* (3(8)), 4–9.

Chaikovskaya, E. (2015). Devising an energy saving technology for a biogas plant as a part of the cogeneration system. *Eastern-European Journal of Enterprise Technologies,* (3/8(75)), 47–53. https://doi.org/10.15587/1729-4061.2015.44252.

Chaikovskaya, E. (2016). Development of energy-saving technology maintaining the functioning of a drying plant as a part of the cogeneration system. *Eastern-European Journal of Enterprise Technologies*, (3/8 (81)), 42-46. https://doi.org/10.15587/ 1729-4061.2016.72540.

Chaikovskaya, E. (2016). The development of energy-saving operation technology of the biodiesel plant as a part of the cogeneration system. *Eastern-European Journal of Enterprise Technologies*. (1/8 (79)), 4-11. https://doi.org/10.15587/1729-4061.2016.59479.

Chaikovskaya, E. (2017). Development of energy-saving technology to support functioning of the lead-acid batteries. *Eastern-European Journal of Enterprise Technologies*, (4/.8 (88)), 56–64. https://doi.org/10.15587/1729-4061.2017.108578.

Chaikovskaya, E. (2018). Development of energy – saving tehnology for maintaning the functioning of heat pump power supply. *Eastern-European Journal of Enterprise Technologies,* (4/.8 (94)), 13–23. https://doi.org/10.15587/1729-4061.2018.139473.

Chaikovskaya, E. (2019). Development of energy-saving technology to maintain the functioning of a wind-solar electrical system. *Eastern-European Journal of Enterprise Technologies,* (4/.8 (100)), 57–68. https://doi.org/10.15587/1729-4061. 2019.174099.

Chaikovskaya, E. (2020). Development of Smart Grid tehnology for maintaning the functioning of a biogas cogeneration system. *Eastern-European Journal of Enterprise Technologies.* (8/3 (105)), 56–68. https://doi.org/10.15587/1729-4061. 2020.205123.

Chaikovskaya, E. (2021). Development of Smart Grid tehnology for maintaning the functioning of Photoelectric charging stations. *Eastern-European Journal of Enterprise Technologies.* (8/3 (111)), 14–24. https://doi.org/10.15587/1729-4061.2021.235120.

Chaikovskaya, E. (2022). Book. Chapter 6. Smart Grid Technology for Maintaining the Functioning of a Wind-Solar Electric System. *Advances in Energy Research. Volume 35. Morena J. Acosta (Editor).* NOVA Science publishers. New Work. P.215-236. Publication Date: November 11, 2021 Pages: 226. ISBN: 978-1-68507-373-1 (e-book) ISSN 2157-1562. https://doi.org/10.52305/EEEQ2450.

Chaikovskaya, E. E. (1999) Sinergeticheski podhod pri razrabotke ekspertnih sistem [*Synergetic approach in the development of expert systems*]. *Proceedings of the Odessa Polytechnic University*. №2(8). 126-128.

Chaikovskaya, E. E. (1999). Dinamicheskaya podsistema kak osnova ekspertnich system [*Dynamic subsystem as the basis of expert systems*]. *Proceedings of the Odessa Polytechnic University*. №3(9). 108-110.

Chaikovskaya, E. E. (2000). Matematicheskoe modelirovanie dinamiki energeticheskih system kak osnovi diagnostiki [*Mathematical modeling of the dynamics of energy systems as the basis for diagnostics*], *Proceedings of the Odessa Polytechnic University*. №3(12). 83-86.

Chaykovskaya, E. E. (2016). Soglasovanie proizvodstva i potrebleniya energii na osnove intellektual'nogo upravleniya teplomassobmennymi protsessami [*Coordination of energy production and consumption based on intelligent control of heat and mass transfer processes*]. *XV Minsk International Forum on Heat and Mass Transfer: Proceedings of XV Minsk International forum on heat and mass transfer. Section 8. Heat and mass transfer in energy processes and equipment. Energy Saving.* Minsk, 1–12.

Chaykovskaya, E. E. (2021). Kompleksnoie modelirovanie protcessov teplomass-opierenosa kak osnova upravlienia na urovnie priniatia rechenii. XVI Minskiy mezhdunarodnyy forum po teplomassobmenu [*Complex modeling of heat and mass transfer processes as a basis for control at the level of speech acceptance. XVI Minsk International Forum on Heat and Mass Transfer*]. *Abstracts of reports and communication of XVI Minsk. International forum on heat and mass transfer. Section 8. Modeling and control of heat and mass transfer processes.* 1290-1293.

Daniyan, I., Daniyan, O., & Abiona, O., Mpofu, K. (2019). Development and Optimization of a Smart System for the Production of Biogas using Poultry and Pig Dung. *Procedia Manufacturing,* 35, 1190–1195.

Davye, M., Daranith, & Ch., Dae-Hyun, Ch. (2020). Sensitivity analysis of volt-VAR optimization to data changes in distribution networks with distributed energy resources. *Applied Energy*, 261,114331.

Dixon, J., Bell, K. (2020). Electric vehicles: Battery capacity, charger power, access to charging and the impacts on distribution networks. *eTransportation*, 100059. doi: https://doi.org/10.1016/j.etran.2020.100059.

Elma, O. (2020). A dynamic charging strategy with hybrid fast charging station for electric vehicles. *Energy*, 117680. https://doi.org/10.1016/j.energy.2020.117680.

Fathabadi, H. (2020). Novel stand-alone, completely autonomous and renewable energy based charging station for charging plug-in hybrid electric vehicles (PHEVs). *Applied Energy*, 260, 114194. https://doi.org/10.1016/j.apenergy.2019.114194.

Ferro, G., Laureri, F., Minciardi, R., Robba, M. (2019). A predictive discrete event approach for the optimal charging of electric vehicles in microgrids. *Control Engineering Practice*, 86, 11–23. https://doi.org/10.1016/j.conengprac.2019.02.004.

Gholizadeh, T., Mohammad V., & Rostamzadeh, H. (2019). Energy and exergy evaluation of a new bi-evaporator electricity/cooling cogeneration system fueled by biogas *Cleaner Production*, 233, 1494–1509.

Guliaev V. A. (1990). Diagnostika entrgeticheskih i elektronnich sistem. [*Diagnostics of energy and electronic systems*]. *Collection of scientific papers*. Kiev. Naukova dumka.

Haken H. (1996). Principles of Brain Functioning *Springer Series in Synergetics.* Berlin: Springer. 67.

Helietuha, H. H. Geleznaia, T. A., Kucheruk P. P. Oleinik, E. N., Triboi A. V. (2015). Bioenergy in Ukraine: current state and prospects of development. Part 2. *Industrial Heat*, 37 (3), 65–73.

Jia, Y., Liu, X. J. (2014). Distributed model predictive control of wind and solar generation system. *Proceedings of the 33rd Chinese Control Conference.* doi: https://doi.org/10.1109/chicc.2014.6896301.

Jiao, Z., Lu, M., Ran, L., Shen, Z.-J. M. (2020). Infrastructure Planning of Photovoltaic Charging Stations. SSRN. *Electronic Journal*. https://doi.org/10.2139/ssrn.3560677.

Jordán, J., Palanca, J., del Val, E., Julian, V., Botti, V. (2021). Localization of charging stations for electric vehicles using genetic algorithms. *Neurocomputing*, 452, 416–423. https://doi.org/10.1016/j.neucom.2019.11.122.

Liu, J., Dai, Q. (2020). Portfolio Optimization of Photovoltaic/Battery Energy Storage/Electric Vehicle Charging Stations with Sustainability Perspective Based on Cumulative Prospect Theory and MOPSO. *Sustainability*, 12 (3), 985. https://doi.org/10.3390/su12030985.

Perera, A., Vahid M., & Wickramasinghe, P., Scartezzini, J. (2019). Redefining energy system flexibility for distributed energy system design. *Applied Energy*, 253, 113572.

Prigozin I. (1960). Vvedenie v termodinamiku neobratimih prosessov [*Introduction to the thermodynamics of irreversible processes*]. *M. Publishing house of foreign literature.* 160.

Rostampour, V., Jaxa-Rozen, M., & Bloemendal, M., Kwakkel, V., Keviczky, T. (2019). Aquifer Thermal Energy Storage (ATES) smart grids: Large-scale seasonal energy storage as a distributed energy management solution. *Applied Energy*, 242, 624–639.

Saad, A., Samy, F., & Osama, M. (2019). A secured distributed control system for future interconnected smart grids. *Applied Energy*, 243, 57–70.

Von Bertalanffy L. *General System Theory, Foundations, Development, Applications:* rev. ed. (1976). N. Y. George Brazilier.

Xiqiao, L., Yukun, L., & Xianhong, B. (2019). Smart grid service evaluation system. *Procedia CIRP*, 83, 440–444.

Yinan, L., Wentao, Y., & Ping, H., Chang, Ch., Xiaonan, W. (2019). Design and management of a distributed hybrid energy system through smart contract and blockchain. *Applied Energy*, 248, 390–405.

Zaher, G. K., Shaaban, M. F., Mokhtar, M., Zeineldin, H. H. (2021). Optimal operation of battery exchange stations for electric vehicles. *Electric Power Systems Research*, 192, 106935. https://doi.org/10.1016/j.epsr.2020.106935.

About the Author

Eugene Chaikovskaya

Affiliation: PhD, Senior Researcher, Associate Professor.

Education: Department of Theoretical, General and Nonconventional Power Engineering, Odessa Polytechnic National University, Ukraine.

Business Address: Shevchenka ave., 1, Odessa, Ukraine, 65044.

Research and Professional Experience:

- Number of publications in Ukrainian publications: 233.
- Number of publications in indexed foreign publications: 12.
- Scopus index: h index: 2; 9 individual publications (2014 - 2022).
- ORCID: http://orcid.org/0000-0002-5663-2707.
- http://www.scopus.com/authid/detail.uri?authorId=57170828500.
- ID: 57170828500.
- https://www.researchgate.net/profile/Eugene-Chaikovskaya.

Publications from the Last 5 Years:

1. Chaikovskaya, E. (2014). Maintaining the relation between production and consumption of electricity and heat at decision-making level. *Eastern-European Journal of Enterprise Technologies* (3(8)), 4–9. Scopus.
2. Chaikovskaya, E. (2015). Devising an energy saving technology for a biogas plant as a part of the cogeneration system. *Eastern-European Journal of Enterprise Technologies,* (3/8 (75)), 47–53. doi: https://doi.org/10.15587/1729-4061.2015.44252. Scopus.
3. Chaikovskaya, E. (2016). The development of energy-saving operation technology of the biodiesel plant as a part of the cogeneration system. *Eastern-European Journal of Enterprise*

Technologies, (1/8 (79)), 4-11. https://doi.org/10.15587/1729-4061.2016.59479. Scopus.

4. Chaikovskaya, E. (2016). Development of energy-saving technology maintaining the functioning of a drying plant as a part of the cogeneration system. *Eastern-European Journal of Enterprise Technologies,* (3/8 (81)), 42-46. https://doi.org/10.15587/1729-4061.2016.72540. Scopus.
5. Chaykovskaya, E. E. (2016). Soglasovanie proizvodstva i potrebleniya energii na osnove intellektual'nogo upravleniya teplomassobmennymi protsessami [*Coordination of energy production and consumption based on intelligent control of heat and mass transfer processes*]. *XV Minsk International Forum on Heat and Mass Transfer: Proceedings of XV Minsk International forum on heat and mass transfer. Section 8. Heat and mass transfer in energy processes and equipment. Energy Saving.* Minsk, 1–12.
6. Chaikovska, Ye. Ye. (2017). Intelectualni systemi pidtrymky functcionuvania energetichnih system na rivni priniattia rihen [*Intelligent systems supporting the functioning of energy systems at the level of priniattia rigen*]. *Enerhetyca. economica, tehnologii, ekologia,* 3, 114–118. https://doi.org/10.20535/1813-5420.3.2017.117377.
7. Chaikovska, Ye. Ye. (2016). Informatsiyni tekhnolohiyi pidtrymky funktsionuvannia enerhetychnykh system na rivni pryiniattia rishen. Informatyka [*Information technologies supporting the functioning of energy systems at the level of initiation of decisions. Informatics*]. *Kultura. Tekhnika: zb. tez dop. IV ukr.-nim. konf.* Odessa, 32–33.
8. Chaikovskaya, E. (2017). Development of energy-saving technology to support functioning of the lead-acid batteries. *Eastern-European Journal of Enterprise Technologies,* (4/8 (88)), 56-64. https://doi.org/10.15587/1729-4061.2017.108578. Scopus.
9. Chaikovskaya, E. (2018). Development of energy -saving technology for maintaning the functioning of heat pump power supply. *Eastern-European Journal of Enterprise Technologies,* (4/8 (94)), 13–23. https://doi.org/10.15587/1729-4061.2018.139473. Sc-opus.
10. Chaikovskaya, E. (2019). Development of energy – saving technology to maintain the functionimg of a wind-solar electrical system. *Eastern-European Journal of Enterprise Technologies,* (4/8 (100)), 57–68. https://doi.org/10.15587/1729-4061.2019.174099. Scopus.

11. Chaikovskaya, E. (2020). Development of Smart Grid technology for maintaning the functioning of a biogas cogeneration system. *Eastern-European Journal of Enterprise Technologies.* (8/3 (105)), 56–68. https://doi.org/10.15587/1729-4061.2020.205123. Scopus.
12. Chaikovskaya, E. (2021). Development of Smart Grid technology to maintaning the functioning of photoelecteic charging stations. *Eastern-European Journal of Enterprise Technologies.* (3/8 (111)), 14–24. https://doi.org/10.15587/1729-4061.2021.235120. Scopus.
13. Chaykovskaya, E. E. (2021). Kompleksnoie modelirovanie protcessov teplomassopierenosa kak osnova upravlienia na urovnie priniatia rechenii. XVI Minskiy mezhdunarodnyy forum po teplomassobmenu [*Complex modeling of heat and mass transfer processes as a basis for control at the level of speech acceptance. XVI Minsk International Forum on Heat and Mass Transfer. XVI Minsk International Forum on Heat and Mass Transfer*]. *Abstracts of reports and communication of XVI Minsk. International forum on heat and mass transfer. Section 8. Modeling and control of heat and mass transfer processes.* 1290-1293.
14. Chaikovskaya, E. (2022). Book. Chapter 6. Smart Grid Technology for Maintaining the Functioning of a Wind-Solar Electric System. Advances in Energy Research. Volume 35. *Morena J. Acosta* (Editor). NOVA Science publishers. New Work. P.215-236. Publication Date: November 11, 2021 Pages: 226. ISBN: 978-1-68507-373-1 (e-book) ISSN 2157-1562. https://doi.org/10.52305/EEEQ2450.

Index

D

E

F

G

H

I

L

M

O

P

R